NCLEX Simplified
by
Lisa Chou

Introduction

Welcome to NCLEX Simplified, a book created entirely out of original material by Lisa Chou to support the students she tutors. The book is composed of three main sections: purely informational material paired with chapter quizzes, focused question sections and extra resources. The material is separated into several sections. Remember that it does not cover EVERYTHING that you need for NCLEX- it only covers the most frequent topics on the test.

I recommend that you visit www.nclexsimplified.com to access the latest information. You may also find further resources to help you study.

I hope that you get a lot of out of this review and are able to pass your test! If you do, please review me on the NCLEX Simplified Facebook page and spread the words to your friends!

Table of Contents

CX: complaint of.

Cardiac

O. Hypo

1. Angina

Stable if lasts less than 5 minutes, pain is same, subsides with rest/nitro
Unstable if severe pain, is not relieved by rest or nitro, different pain location/quality

Nitroglycerin	Vasodilator *Give 3x, 5 min apart for angina*	S/E: Flushing, hypotension, headache
Sildenafil	Tx of erectile dysfunction (Viagra) Vasodilator	S/E: Flushing, hypotension, priapism DO NOT USE WITH NITROGLYCERIN

2. Hypertension

Essential/Primary - HTN with no identifiable cause, most common type (95%), genetics + environment

Secondary- HTN caused by something identifiable. Ex: endocrine diseases, kidney diseases, tumors, side effects of medications, etc.

-risk factors for HTN can be either modifiable or non-modifiable

non-modifiable- age, gender, genetic factors(family hx), race

modifiable- obesity, high sodium intake, low potassium intake, low Vitamin D, alcohol consumption, reduced physical activity, high levels of stress

-associated with increased risk of stroke/TIA (transient ischemic attack) in which clots or bleeds/ruptures block blood flow to the brain

Verapamil, Cardizem	Calcium channel blockers Treats HTN	S/E: Orthostatic hypotension, angina, impotence, avoid grapefruit juice
Lisinopril (CCB) (Prinvil)	Anti HTN, ace inhibitor	S/E: Orthostatic hypotension, persistent dry cough
Propranolol (BB) (Indrel)	Anti HTN, beta blocker Contraindicated for patient with chronic respiratory issue	S/E: Orthostatic hypotension, bradycardia, dysrhythmias, diabetics have higher risk of hypoglycemia, BRONCHOSPASM
Losartan (Cozaar)	Anti HTN, ace receptor blockers	S/E: Orthostatic hypotension, monitor for edema and renal failure
Methyldopa	Anti HTN	S/E: Orthostatic hypotension, do not d/c abruptly
Cholestyramine (Questran)	Antilipemic (anti-cholesterol)	S/E: Fat-soluble vitamin deficiency Sprinkle powder in beverage or wet food

4

3. Stroke/ Transient Ischemic Attack (TIA)

Bell's palsy

s/s: slurring of words, numbing or tingling on one side of the body, facial droop, confusion, trouble finding words

FAST- facial droop, arm weakness, slurring of words, time to call 911

-can be either hemorrhagic or ischemic

        -ischemic- cause is clot, treated with tPA if within 12 hours and has no risk factors

        -hemorrhagic- cause is a bleed, more complex; NOT treated with tPA

TIA - "mini strokes", transient block of blood flow that can easily recur - important to have pt seek medical attention if symptoms recur because that can lead to stroke

Nursing: important to know what TIME symptoms started

-ASA should not be given until type of stroke is known

-Pts have increased risk for aspiration and need a swallow evaluation

-in immediate recovery, pt is at risk for increased ICP

-over the months following stroke, pt's level of functioning may get better or worse day-to-day

tPA (tissue plasminogen activator)	Protein that breaks down blood clots	S/E: bleeding, hemorrhage Contraindicated: Recent head injury, bleeding issues, ulcers, recent surgery, severe HTN or previous stroke within 2 months -given in first 12 hours after MI and non-hemorrhagic stroke

4. Arterial v. Venous Insufficiency

Arterial:

S/s: Cool shiny skin, cyanosis, pale limbs, ulcers, gangrene, paresthesias

-intermittent claudication- pain when walking/moving, relieved by rest

Nursing: monitor peripheral pulses, good foot care, do not cross legs, regular exercises, place legs in DEPENDENT position to relieve pain at rest

↓ *dangle legs: decrease return of blood & reduce pulmon congestion.*

Venous:

S/s: aching, cramping pain, thickened and toughened brown skin, ulcers, varicose veins, edema

Nursing: elastic support stockings, compression bandages for wound healing, anticoagulation therapy (see below), ELEVATE legs to relieve pain/swelling

5. Deep Vein Thrombosis *(blood clots in legs)*

Risks- pregnancy, oral contraceptives, smoking, surgery, bedrest, airplane/travel

prevent with elastic stockings

S/s: localized edema of one extremity, calf pain, warm skin, redness, + Homan's sign

Tx: bedrest, elevate extremity, anticoagulants

Heparin	Anticoagulant	S/E: Thrombocytopenia, anemia
		Monitor INR, PTT (partial thromboplastin time)
		Antidote: Protamine sulfate
Warfarin	Anticoagulant	S/E: Diarrhea, rash
		Monitor INR, PT (prothrombin time), avoid diet high in Vitamin K
		Antidote: Vitamin K
Clopidogrel	Antiplatelet	S/E: Hemorrhage
Acetylsalicylic Acid (ASA) *aspirin*	Antiplatelet	S/E: GI bleeding, metabolic acidosis, tinnitus, do NOT give to children

6. Cardiac Catheterization

def: catheter is threaded through the femoral artery and dye is injected to observe blood flow

Check for shellfish or iodine allergies- will be allergic to dye

normal dye reactions: flushing in the face, burning/warmth at IV site, taste of salt

Pt lies on his back for several hours after the procedure

Must keep affected leg straight, keep pressure bandage on, check pulses in each leg

Watch for hemorrhage, check pulse and vitals

7. CABG (Coronary Artery Bypass Graft)

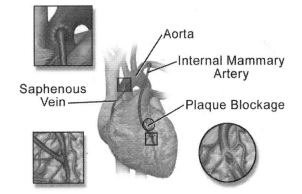

(picture credit: study.com)

def: open-heart surgery where a vein is taken from the leg and grafted around a blockage in the heart

-involves breaking ribs, ventilator, bypass machine, weeks to months of recovery

-vein site may weep fluid for up to one week

-possible cx:

cardiac tamponade:

-s/s: sinus tachycardia, hypotension, muffled heart sounds, enlarged cardiac silhouette on CXR

↑ chest - xray

8. Heart failure

Left-sided: Dyspnea, cough, orthopnea, edema and weight gain

Right-sided: (aka cor pulmonale) initially dependent edema, jugular vein distention, hepatomegaly (JVD)

Take a daily weight at the same time each day, report gain of >2 lbs

Diet: sodium restricted

Angina
R-Renexin
N: nitrates
S: Statins
(preventitive)

Digitalis	Cardiac glycoside	(see Afib above)
Hydrochlorothiazide	Diuretic (K+ wasting) Do not give at bedtime, weigh pt daily, encourage K+ foods	S/E: Hypokalemia, hyperglycemia, hypotension
Furosemide	Diuretic (K+ wasting) Do not give at bedtime, weigh pt daily, encourage K+ foods	S/E: Hypokalemia, hyperglycemia, hypotension
Potassium chloride	K+ supplement Monitor EKG, take after meals	S/E: Dysrhythmias, cardiac arrest, resp paralysis
Triamterene	Diuretic (K+ sparing) Avoid salt subs, give with meals	S/E: Hyperkalemia, hyponatremia
Spironolactone	Diuretic (K+ sparing) Avoid salt subs, give with meals	S/E: Hyperkalemia, hyponatremia, hepatic/renal damage

(Hyper K)
HF = MURDER

Hypokalemia- abdominal cramping, muscle weakness, lethargy (LAM)

Hyperkalemia- MURDER - muscle weakness, urine (oliguria/anuria), respiratory depression, decreased cardiac function, EKGs, reflexes

9. Myocardial Infarction (MI) (Troponin - protien) indicator of damage to the ♡ •measured in the blood to know diff btwn MI or angina.

s/s: SOB, chest/arm pain are hallmark signs, but may simply include nausea and diaphoresis

*women-specific symptoms: dizziness, indigestion, pain between shoulder blades, jaw, or back, and fatigue x several days, insidious onset

EKG- ST segment elevation, T wave inversion

MONA
morph
O_2
NTG
Asp.

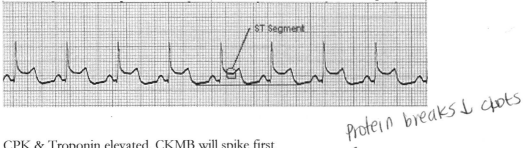

Protein breaks ↓ clots

CPK & Troponin elevated, CKMB will spike first

Tx: MONA (morphine, oxygen, nitroglycerin, ASA), possibly tPA within 12 hours, monitor EKG, semi-Fowler's position, bedrest, recovery through cardiac rehab

Cx: Fatal dysrhythmias

Morphine	Opioid	S/E: Sedation, resp depression, constipation
		*Monitor vitals, bowel patterns, observe for dependence

10. Shock

S/s: Tachycardia, decrease in urinary output, anxiety, weak pulse, decrease in blood pressure, cool clammy skin, nausea and vomiting

possible specific types: cardiogenic, hypovolemic

POSITION: elevate extremities, knees straight, head slightly elevated

GIVE: IV fluids, vasoconstrictors, blood if blood loss occurred

Norepinephrine/ epinephrine	Vasoconstrictor, increases BP and cardiac output	S/E: Tissue necrosis with extravasation
Dopamine	Used in shock in low doses in order to dilate the renal/coronary arteries	Headache = sign of drug excess

11. Pacemaker guidelines

No raising arm above head for one week after surgery, count the pulse at least once per week, no vacuuming, instruct client to carry ID, can go through airport security, no airport hand scanning, no MRI

Cx: dizziness, hiccups, chest pain, SOB, increase or decrease in heart rate, pulse below 60

12. Pulmonary artery catheter

def: catheter (aka Swan-Ganz line) inserted into the pulmonary artery that measures cardiac output and pressure in left ventricle

-allows for pulmonary capillary wedge pressure (balloon from catheter is inflated and floats to be "wedged" inside the small pulmonary arterial branch)

-if PCWP if elevated, suggests failure of left ventricle - normal PCWP = 2-15

13. Central Venous Pressure

def: placed in superior vena cava, looks at pressure in right atrium

-measurement determined by blood volume, vascular tone, action of right side of the heart

-fluid will fluctuate with respirations

-normal = 4-10cm/H_2O.

>10 = hypervolemia, CHF, pericarditis

<4 = hypovolemia

14. Arterial Line — vein in wrist by thumb.

def: catheter used for measuring BP, HR, frequent blood draws, ABGs, NEVER use for meds

15. Valve Auscultation Locations

	PMI location- 5th ICS space, midclavicular APE To Man Aortic valve area: 2nd ICS, RSB Pulmonic valve area: 2nd ICS, LSB Erb's point: 3rd ICS, LSB Tricuspid valve area: 5th ICS, LSB Mitral valve area: 5th ICS, midclavicular

16. Rheumatic fever

- causes heart valve damage with group A beta-hemolytic streptococcal infection

-take abx prior to dental appts, surgery

17. Orthostatic Vital Signs

Take a blood pressure lying down, sitting, standing up to check blood pressure changes

18. Abdominal and Thoracic Aortic Aneurysms

Causes: atherosclerosis, congenital weakness, chronic infection

s/s: mass in the abdomen, pulsating sensation in the abdomen, blood pressure may be lower in legs than arms or in the left arm and legs than in the right arm

Nrsg: avoid palpation of the abdominal wall, monitor blood pressure frequently, monitor renal function

cx: ecchymoses of the scrotum, perineum, or expanding hematoma = graft leak

cx: severe back pain and bilateral flank ecchymoses, sudden loss of consciousness, shock = rupture and hemorrhage tx: insert IV immediately for fluid/blood pressure replacement, maintain blood pressure

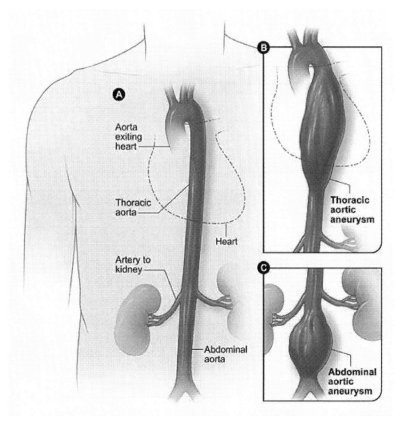

(picture credit: http://omnimedicalsearch.com)

Cardiac Review Quiz

1. Biggest side effect of ace inhibitors?

2. Treatment for angina?

3. Arterial insufficiency should be placed in what position to relieve pain?

4. Post cardiac catheter position for the patient?

5. What does an elevated ST segment indicate?

6. Which is an emergency: Cardioversion or defibrillation.

7. Treatment for DVT?

8. Signs and symptoms of right-sided heart failure?

9. What are the valves in APE to MAN and their locations?

10. An EKG from a patient with an MI shows...

11. Cardiac enzymes?

12. Shock treatment position?

13. Explain how to perform orthostatic vital signs.

14. What are the signs and symptoms of cardiac tamponade?

15. What are the signs and symptoms of an abdominal or thoracic aortic aneurysm?

16. What is a normal PCWP?

NCLEX Simplified CORE Questions: Cardiac

1. The nurse works on a medical/surgical unit and cares for a patient receiving Lanoxin (Digoxin) and Furosemide (Lasix). The nurse knows that which of the following, if reported by the patient, must be assessed IMMEDIATELY?

A: Night sweats and headache.

B: Vomiting and halos around lights. ✓

C: Stomach upset and headache.

D: Low blood pressure and dark urine.

2. After a cardiac catheterization, the nurse instructs the patient to lie in which of the following positions?

A: Bedrest, supine. ✓

B: Prone.

C: Left side-lying.

D: Right side-lying.

3. A patient comes to the clinic complaining of a pulsating sensation in her stomach. The nurse puts her stethoscope over the patient's stomach and hears a steady pulse. The patient's blood pressure is 100/50 in her left arm and 125/65 in her right arm. The nurse suspects the patient is suffering from which of the following conditions?

A: Hypertension

B: Aortic aneurysm ✓

C. Stroke

D. DVT

4. A patient with venous stasis ulcers complains of pain and swelling. The patient refuses to prop their legs up on an ottoman and asks what else she can do to help the ulcers heal. Which of the following suggestions could the nurse make?

A: "Why don't you try getting a massage for your legs every now and then? It would do them good."

B: "You should wear compression stockings."

C: "There is no other option. You really have to elevate them, even if it's uncomfortable at first."

D: "A diuretic prescribed by the doctor can help your symptoms."

5. Which is the primary consideration when preparing to administer thrombolytic therapy to a patient who is experiencing an acute myocardial infarction (MI)? *when did the pain start.*

A: History of heart disease.

B: Sensitivity to aspirin.

C: Size and location of the MI.

D: Time since onset of symptoms. ✓

6. Which of the following diagnostic studies most likely would confirm a myocardial infarction?

A: Serum myoglobin level

B: Creatinine kinase (CK)

C: Troponin T levels ✓

D: White blood cell count (WBC)

7. Which of the following prescriptions would the nurse expect to be ordered for a patient with atrial fibrillation?

A: Simvastatin

B: Aspirin

C: Warfarin ✓

D: Vancomycin

8. An elderly client is being monitored for evidence of congestive heart failure. To detect early signs of heart failure, the nurse would instruct the certified nursing attendant (CNA) to do which of the following during care of the patient?

A: Monitor vital signs every 15 minutes and report each reading to the nurse.

B: Assist the client with ambulation three times during the shift.

C: Observe EKG readings and report abnormalities to the nurse.

D: Accurately weigh the patient and report and record the readings. *Fluid retention*

9. A patient has been diagnosed with essential hypertension. The nurse knows that this type of hypertension: *2nd has something identifiable*

A: is unpreventable.

B: has modifiable risk factors.

C: has no identifiable cause. ✓

D: is secondary to a disease or condition.

10. Sildenafil should never be combined with:

A: Aspirin
B: Nitroglycerin
C: Metoprolol
D: Nitrous

11. A patient in the ICU goes into ventricular tachycardia. With a team of healthcare professionals, the nurse performs CPR. The nurse knows she is giving effective compressions if the patient's sternum is compressed..
.

A: 1/3 of the chest.
B: 1/8 of the chest.
C: 1.5 inches.
D: 2 inches.

12. A patient has a history of rheumatic fever. Which of the following precautions is indicated for this patient to take?

A: Handwashing.
B: Antibiotics prior to dental work.
C: Lifelong antibiotics.
D: Antibiotics prior to surgery only.

13. Twelve hours after a patient returns from a CABG, the nurse finds the patient short of breath and complaining of chest pressure. What should the nurse suspect?

A: Myocardial infarction
B: Cardiac tamponade
C: Cardiac arrest
D: Airway irritation related to surgery

14. A patient is diagnosed with a DVT. Which of the following, if stated by a UAP (Unlicensed Assistive Personnel), requires intervention by the nurse?

A: "I am going to ambulate the patient to the bathroom and assist her with using the toilet."
B: "Sometimes I wish the patient would talk a little less. It's hard to get out of the room."
C: "The patient states that her leg is warm, swollen, and painful to the touch."
D: "The patient and I were talking, and we both play softball!"

15. A patient has been taking a heavy aspirin regimen for the past two months. Which side effects, if noted by the patient, are directly related to overdose of aspirin?

A: Confusion

B: Clay-colored stools

C: Tinnitus

D: Diarrhea

16. The nurse enters a patient's room and observes via the monitor that their vital signs are as follows: blood pressure 80/42, heart rate 118, respirations 32. The nurse will place the patient in which of the following positions?

A: Trendelenburg position.

B: Head slightly elevated and midline, knees straight, all four extremities elevated on pillows.

C: HOB elevated 30 degrees, body midline.

D: Feet elevated only.

17. A patient at the outpatient clinic describes leg pain that occurs while walking and is relieved with rest. The patient asks why this happens to him. The nurse knows that the pain is related to:

A: Venous insufficiency.

B: Heart failure.

C: Arterial insufficiency.

D: Angina.

18. The nurse in the outpatient care clinic cares for a client diagnosed with heart failure. Which of the following orders, if written by the physician, should the nurse question?

A: "Administer 0.9% NS solution IV at a rate of 125mL/hr."

B: "Administer Lasix 40mg twice daily PO."

C: "Administer Potassium 40mEq tab once daily PO."

D: "Administer Lactated Ringers solution IV at a rate of 50mL/hr."

19. A patient is scheduled for a cardiac catheterization this afternoon. Which of the following, if noted in the patient's chart by the nurse, is a contraindication to the test?

A: The patient has a history of asthma.

B: The patient has a history of schizophrenia.

C: The patient has an allergy to eggs.

D: The patient has an allergy to clams.

20. Shortly after a radial cardiac catheterization, a patient has bounding radial pulses bilaterally, with compression device intact. With the next neurovascular check, the patient complains of numbness and pain in the right hand. The cardiac/vascular nurse notes a diminished pulse, and a cool and cyanotic hand. The nurse's next intervention is to:

A: perform an Allen's test.

B: call the physician.

C: reduce the pressure on the puncture site.

D: use the Doppler to assess for pulse signals.

21. The nurse prepares the client for insertion of a pulmonary artery catheter (Swan-Ganz catheter). The nurse teaches the client that the catheter will be inserted to provide information about:

A: Stroke volume

B: Cardiac output

C: Venous pressure

D: Left ventricular function

22. What is the most important nursing action when measuring a pulmonary capillary wedge pressure (PCWP)?

A: Have the client bear down when measuring the PCWP

B: Deflate the balloon as soon as the PCWP is measured

C: Place the client in a supine position before measuring the PCWP

D: Flush the catheter with heparin solution after the PCWP is determined.

23. The healthcare provider is performing an assessment on a patient who is taking propranolol (Inderal) for supraventricular tachycardia. Which assessment finding is an indication the patient is experiencing an adverse effect of this drug?

A: Dry mouth

B: Bradycardia

C: Urinary retention

D: Paresthesias

24. A patient who is in cardiogenic shock has a urine output of 20mL/hr. When further assessing the patient's renal function, what additional findings are anticipated?

A: Decreased urine specific gravity.

B: Increased blood urea nitrogen (BUN)

C: Increased urine sodium

D: Decreased serum creatinine

25. The healthcare provider is explaining to a patient the reason why vasculitis causes damage to the blood vessels in the body. Which of the following is the best explanation?

A: "Atherosclerotic plaque is causing narrowing of your blood vessels."

B: "High blood pressure is damaging the delicate lining of your arteries."

C: "Your immune system is overreacting and causing damage to the vessel walls."

D: "Your platelets are extra sticky so they form lots of clots in the blood vessels."

Answers to CORE Questions: Cardiac

1. Answer: B: Lasix causes the patient to lose potassium. Digoxin, if taken with a low potassium level, can become toxic and show signs/symptoms of nausea, vomiting, and halos around lights.
2. Answer: A: After a cardiac catheterization, the patient needs to lie still and flat. The most often-used site to thread the catheter is the femoral artery, which should have a pressure bandage applied. This site should be kept as still as possible and should not be bent/moved if possible.
3. Answer B: Aortic aneurysms may cause differing blood pressures between arms or legs. Abdominal aortic aneurysms may cause a pulsating sensation in the stomach or appear as a mass with a visible pulse.
4. Answer: B: Compression stockings will encourage better drainage from the legs and improve the stasis ulcers. The legs should never be massaged r/t DVT concerns.
5. Answer: D: The time since the onset of symptoms is critical because thrombolytic therapy cannot be used once over 12 hours since onset has passed.
6. Answer: C: Troponin levels are specific to cardiac muscle enzymes while the other tests listed are non-specific muscle enzymes.
7. Answer: C: Warfarin is a blood thinner, which would be ordered for a patient in atrial fibrillation to prevent clots.
8. Answer: D: Weighing the patient will assess fluid status. Sudden gain of 2-3 lbs in a 24 hour period is an early indication of worsening heart failure.
9. Answer: C: Essential hypertension has no identifiable cause. It may have either modifiable or non-modifiable risk factors.
10. Answer: B: Sildenafil (also known as Viagra) should never be combined with Nitroglycerin because, together, they can cause the patient's blood pressure to become too low. (Both medications are vasodilators.)
11. Answer: A: The adult patient's chest should be compressed to ⅓ the depth of the anterior chest. An infant's chest should be compressed two inches during CPR.
12. Answer: B: A patient with a history of rheumatic fever may have heart valve damage and should take antibiotics prior to both dental work and surgery.
13. Answer: B: A patient who received a CABG procedure is at risk for cardiac tamponade, where blood pools around the heart inside the pericardial sac and makes it difficult for the heart to beat.
14. Answer: A: The patient diagnosed with DVT should not walk to the bathroom because s/he risks dislodging the clot. A leg that is swollen, painful, and warm is typical of a DVT.
15. Answer: C: Tinnitus is a common side effect of long-term or heavy aspirin usage.
16. Answer: B: The patient described is going into shock and should be placed in the NCLEX-approved shock position: body midline and center with all extremities elevated on pillows.
17. Answer: C: The occurrences that the patient describes are known as intermittent claudication and occurs only with arterial insufficiency.
18. Answer: A: A rate of 125mL/hr is too high for a heart failure patient and will increase preload and afterload, as well as cause the patient to develop edema and possibly crackles.
19. Answer: D: A patient who is allergic to clams may not receive the dye used during a cardiac catheterization.
20. Answer: B: Call the physician. There is more than enough evidence (diminished pulse, cool and cyanotic hand) to support suspicion of a clot, and it is imperative to call the physician right away.
21. Answer: D: The catheter is placed in the pulmonary artery. Information regarding

left ventricular function is obtained when the catheter balloon is inflated.

22. Answer: B: In order to measure the PCWP, the balloon must be inflated and temporarily block blood flow. It should therefore be deflated immediately following obtainment of an accurate measurement.

23. Answer: B. As a beta blocker, propranolol works to prevent an arrhythmia like SVT from occurring by slowing down the heart rate. One side effect is that the medication slows down the heart rate excessively- bradycardia.

24. Answer: B: The BUN will also increase because it indicates kidney strain.

25. Answer: C: Vasculitis is an inflammation of the veins that causes damage to the blood vessels of the body through repeated flare-ups.

Want more questions like this? NCLEX Simplified believes that you don't have to practice an endless amount of questions- just the right ones that will teach you what you need to know. Visit **NclexSimplified.com** to find the rest of the questions and study smarter, not harder!

EKG Rhythms & Arrhythmias

1. Cardiac Conduction System

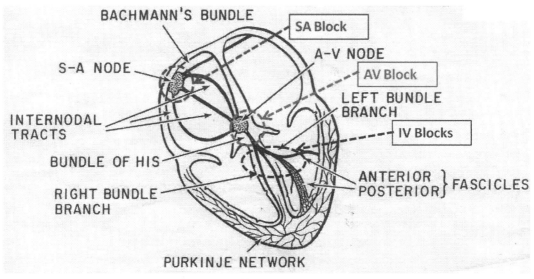

(picture credit: http://omnimedicalsearch.com)

-impulses go through SA node first, then AV node, then IV blocks to the Purkinje fibers

-SA node releases impulses more rapidly, therefore assuming control of process

-AV node generates impulses when SA fails, rate of 40-60bpm

2. Heart Rate

Bradycardia < 60 to 100 (NSR) > Tachycardia

<u>To calculate regular rate:</u> count the number of small boxes between the 'r' waves (top of the qRs) and divide 1500 by that number.

Ex: 12.5 small boxes between 'r' waves means you divide 1500 by 12.5 (rate of 120 bpm)

<u>For irregular heartbeat:</u> count the number of QRS complexes in a 1-minute time interval

3. Cardiac Rhythm Strips

Normal Sinus Rhythm

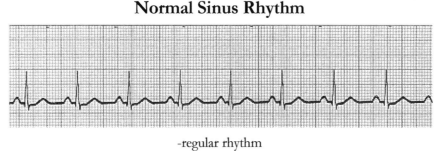

-regular rhythm

-rate between 60 to 100 bpm

-one p for every qrs complex

Bradycardia - <60bpm

tx: atropine, epinephrine, dopamine

Tachycardia - >100bpm

tx: adenosine, diltiazem, beta blockers, amiodarone, digoxin

Atrial Fibrillation

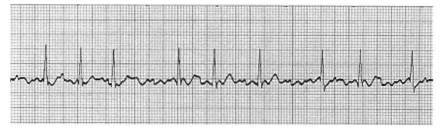

-irregular rhythm

-rate varies from 40 to 120s

-no p waves

Digitalis	Cardiac glycoside Monitor renal func, take apical pulse before admin dose Normal levels: 1.0 to 2.0 in blood	S/E: Bradycardia, visual disturbances, abdominal pain, fine tremors *Works based off K+, so if on a diuretic, watch out!* *Toxic S/E: Halos around dark objects, coarse tremors, nausea, vomiting, diarrhea, heart block, dysrhythmias -can reverse toxicity with Digibind
Rivaroxaban	Anticoagulant (Xarelto)	S/E: bleeding, bruising
Apixaban	Anticoagulant (Eliquis)	S/E: bleeding, bruising

Atrial Flutter

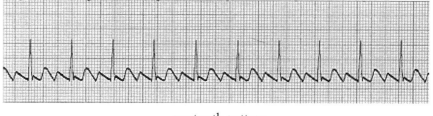

-sawtooth pattern

-multiple p waves for each qrs complex

Supraventricular Tachycardia (SVT)

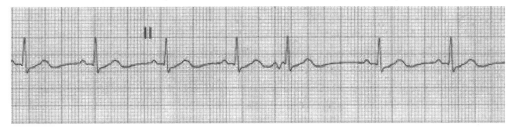

-electrical activity that originates from the upper chambers of the heart and interferes with the signal from the SA node

s/s: dizziness, palpitations, rate >200

tx: bear down, give antiarrhythmic

Adenosine	Antiarrhythmic, treats abnormal heart rhythms like SVT, Wolf-Parkinson-White Syndrome, etc.	S/E: Chest discomfort, lightheadedness, dizziness, throat or jaw discomfort
	*when pushed rapidly, causes asystole in order to let SA node resume sinus rhythm	

PACs (Premature Atrial Contractions)

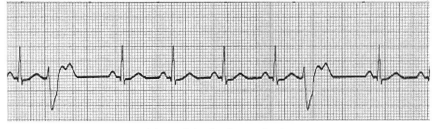

-atria contract too soon

-appear in completely healthy hearts

-medically unconcerning

PVCs (Premature Ventricular Contractions)

-"hiccup" in the heart rhythm

-appear in completely healthy hearts

-only a problem if >6 per minute

Ventricular Tachycardia

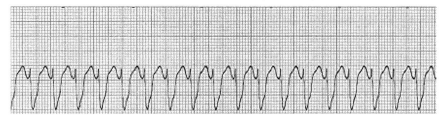

-repeated pattern

-rate >200

-no p waves, only qrs complexes

-heart needs to be shocked out of this rhythm

tx: epinephrine, amiodarone, lidocaine

Ventricular Fibrillation

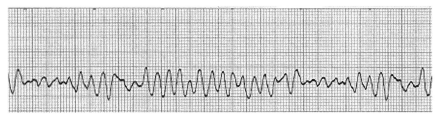

-unorganized, no pattern

-no rate

-heart needs to be shocked out of this rhythm

tx: epinephrine, amiodarone, lidocaine

Aystole

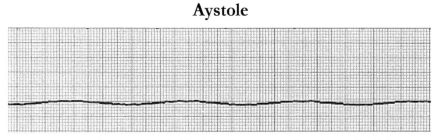

- no rhythm. Unshockable. Needs compressions immediately!

tx: epinephrine

4. Cardioversion/Defibrillation

Defibrillation- pt is unconscious, no stable rhythm- immediate shock

-possible rhythms: ventricular fibrillation, ventricular tachycardia

Cardioversion- while pt is awake and stable- signed consent, no food/fluids for 6 hours before, administer sedation, push sync button to match machine with rhythm, shocks patient into asystole to allow natural pacemaker (SA node) to take over

-possible rhythms: atrial fibrillation, SVT, atrial flutter

Xylocaine	Antidysrhythmic (Lidocaine)	S/E: Slurred speech, confusion, bradycardia
Cordarone	Antidysrhythmic	S/E: Liver toxicity
Atropine sulfate	Tx of bradycardia, sometimes used to dry secretions in mouth	S/E: Headache, dry mouth

5. CPR (Cardiopulmonary Resuscitation)

1. Ask the client if they are okay (check for consciousness)

2. If no response, activate emergency services and request AED.

3. Check for breathing.

4. If unsure, check for pulse for no longer than 10 seconds.

5. If none (or weak), begin 30 compressions that compress the sternum 2 inches, at a rate of 100 per minute (for infants: 1.5 inches (1/3 of chest))

6. Deliver 2 breaths

7. Repeat until another rescuer or EMS arrives

6. Heart Blocks (1st, 2nd, 3rd)

-a patient who shows a heart block needs a next level of care: ICU, external cardiac pacemaker

First Degree Heart Block

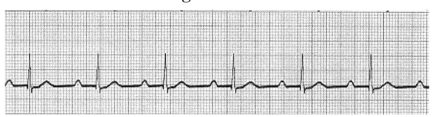

-def: when the electrical signal is slowed through the AV node, demonstrated by PR > .20 sec

Second Degree Heart Block (Type 1 aka Wenckebach)

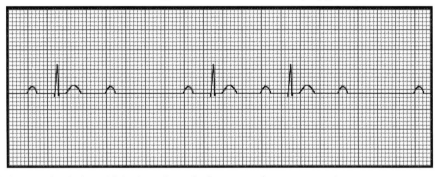

-def: arrhythmia where PR grows longer and longer before dropping the QRS completely

- "long, longer, drop, then you have a Wenckebach"

Second Degree Heart Block (Type 2)

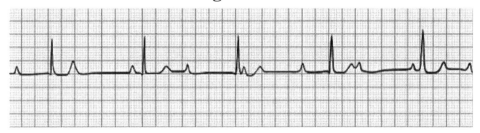

-def: arrhythmia in which the SA node fires, creating a p wave but no QRS response

Third Degree Heart Block

-def: arrhythmia in which there is a completely AV node block, meaning the atria do not communicate or coordinate with the ventricles and vice versa. Each has their own completely

independent rhythm.

Tx: transcutaneous pacing until a permanent pacemaker can be inserted

EKG Review Quiz

1. Atrial fibrillation does/does not have p waves?

2. What rhythms need an immediate shock?

3. Atrial fibrillation puts the patient at risk for...

4. What is the first thing to do for a patient in SVT?

5. What is the treatment for asystole?

Please visit http://EKGsonNCLEX.com for further questions to test your EKG arrhythmia skills.

Circulation Disorders

1. Iron-deficiency Anemia

<u>s/s</u>: dyspnea, palpitations, chronic fatigue, pale, dizziness, weakness

<u>nrg</u>: provide specific treatment, encourage frequent rest periods, provide diet high in protein, iron, and vitamins, provide good oral hygiene, provide oxygen if necessary, administer iron supplements

<u>Note</u>: Iron supplements turn stool black. They should be sipped through a straw to avoid staining teeth.

-If administered IM, use 20 gauge needle and Z-track method into the dorsogluteal muscle

*Pernicious anemia- same symptoms, have no intrinsic factor and intestine cannot absorb Vitamin B12, which is needed to make RBC- need monthly B12 inj for life

Ferrous sulfate	Iron supplement Use straw to avoid staining teeth	S/E: Black stools, nausea, constipation
Deferoxamine	Treats iron toxicity	S/E: red urine
Vitamin B12	Supplement Cobalamin = brand name	S/E: Headache, nausea, upset stomach

2. Blood Transfusions

<u>guidelines</u>: only run with NSS, start blood within 30 minutes of receiving from blood bank, use micron mesh filter, run a 5mL/min x 15 mins, max run time: 4 hours

Reactions Types:

<u>Hemolytic</u>- destruction of red blood cells due to antibodies
s/s: fever, chills, chest pain, back pain, hemorrhage, increased heart rate, shortness of breath, rapid drop in blood pressure, and shock

<u>Allergic</u>- Activates histamines
s/s: wheezing, facial edema

<u>Febrile</u>- Leukocyte incompatibility
s/s: chest tightness, chills, facial flushing, fever up to 104 degrees

Tx for all: Immediately stop transfusion but keep NSS running to preserve IV site.

3. Phlebitis vs Infiltration vs Extravasation

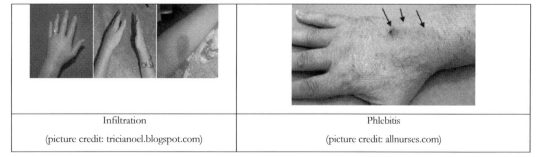

Infiltration	Phlebitis
(picture credit: tricianoel.blogspot.com)	(picture credit: allnurses.com)

Phlebitis- where vein becomes inflamed

s/s: red, tenderness near site, edema

tx: remove IV, elevate limb, apply warm compresses

Infiltration- where IV fluid or medication (non-vesicant) leaks under the skin because IV is no longer in the vein
s/s: cool, edematous site, slowed IV rate

tx: remove IV, elevate limb

Extravasation- where a vesicant (cancer drug, epinephrine, etc.) leaks under the skin.

-first, attempt to remove any residual drug with 1-3mL syringe, cold compresses, then elevate arm x 48 hours

Most stable IV site = distal forearm

-when starting an IV, look lowest (distal) before moving up higher (proximal)

4. Raynaud's Syndrome

-excessively reduced blood flow to extremities, especially fingers, toes, nose

-women are most likely than to men to experience it

s/s: cyanosis in fingers/toes, numbness, tingling, pain

tx: vasodilators, in severe cases: sympathectomy, Botox

Botulinum Toxin Type A	"Botox"- indicated for nerve paralysis, cosmetic sx, migraines, etc.	S/E: Allergic rx, headache, neck/back pain, nausea, diarrhea

5. Burns

First Degree: only epidermis, red and painful skin, = to sunburn without blisters
Second Degree:

        Partial – epidermis and part of dermis, pink/red and wet appearance, blisters

        Full – epidermis and most of dermis, red or white in appearance, dry, needs grafts
Third Degree: All layers destroyed, appears black/white dry

First Degree Burn	Second Degree Burn (Partial)	Second Degree Burn (Full)	Third Degree Burn
venngage	chattycats imgur	slideshare.net	Wikimedia commons

Rule of Nines- to estimate body surface percentage

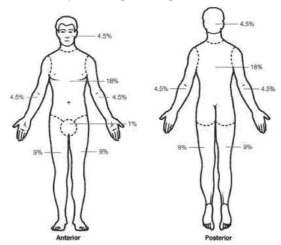

-correct healing = reddened and warm

-if "tender point"= bacterial infection

-hematoma, dehiscence are not good

diet: high vitamin C, high protein, high carbohydrates

Parkland formula- calculates fluid required for first 24 hrs after burns

      -volume = 4 x mass in kg x (Area x 100)

      -from total→ give first half in 8 hours, remaining in next 16 hours

Silver sulfadiazine cream	Tx for burns	S/E: Burning sensation when applied

6. Wounds and Ulcerations

goals: prevent infection, at risk for fluid imbalance

diet: high protein, high vitamin C, high carb

-may use a drainage evacuator, where the exterior bulb is compressed to suction out fluids

Pressure ulcer stages

       1: Intact skin, red – tx: turning, pressure-relieving devices

       2: Partial-thickness loss, superficial ulcer – tx: saline, occlusive dressing

       3: Full-thickness skin loss, deep crater—tx: debridement, wet-to-dry dressings

       4: Full-thickness loss with extensive destruction and tissue necrosis, sinus tracts may be present – tx: nonadherent dressing, changed q8-12hr, skin grafts

Wet-to-dry dressing – purpose is debridement

-removes necrotic tissue by sticking to it as it is removed

-to replace, moisten gauze, wring out, place single layer directly on wound surface, cover with sterile dry gauze and topper dressing

7. Antibiotics to Treat Any and All Infections

-some abx require a peak/trough level

       trough- 30 min prior to giving infusion

       peak- 60 min after the infusion is COMPLETE

Gentamycin, Neomycin	Aminoglycoside Antibiotic	S/E: Ototoxicity, nephrotoxicity
Cephalexin, Ceftriaxone	Cephalosporin Antibiotic	S/E: Superinfections, take with food, avoid alcohol CROSS ALLERGY: PCN
Vancomycin	Glycopeptide, Antibiotic	S/E: Flushing, dizziness, muscle pain/spasms of the chest/back Adverse: Red Man Syndrome-need to do peaks/troughs on it trough: 10-20mcg/mL, peak 20-40mcg/mL
Ciprofloxacin, Levofloxacin	Fluoroquinolones, Antibiotic	S/E: Superinfections, liver toxicity
Penicillin G	Penicillins Antibiotic	S/E: Gastritis, diarrhea, allergy
Sulfamethoxazole, Trimethoprim	Sulfonamides Antibiotic	S/E: Peripheral neuropathy, crystalluria, photosensitivity
Doxycycline	Tetracyclines Antibiotic	S/E: Glossitis, phototoxic reaction, decr oral contraceptives effectiveness
Bacitracin	Antibacterial ointment	S/E: Ototoxicity, nephrotoxicity

9. Specialty Adverse Effects

<u>Red Man Syndrome</u> - occurs due to infusing Vancomycin too fast

nsg: call MD to get order to reduce rate

<u>Stevens-Johnson Syndrome</u> - reaction to a medication or infection that causes flu-like symptoms, painful red rash (maculopapular), skin sloughs

most common med categories: antibiotics, antiepileptics, Allopurinol

Circulation Disorders Quiz

1. What are the two nursing goals of wound care?

2. A patient with iron-deficiency anemia should eat a diet high in…?

3. Describe how to apply a wet-to-dry dressing.

4. Describe the symptoms of Raynaud's syndrome.

5. Signs and symptoms of a hemolytic blood transfusion reaction?

6. What is the difference between infiltration and phlebitis?

7. Correct healing after a burn would be indicated by what appearance?

8. What are some key parts of discharge teaching about iron supplements?

9. What are signs and symptoms of a hemolytic reaction?

10. What is the treatment for a blood transfusion reaction?

Musculoskeletal

1. Fractures

Greenstick fractures- happen in kids at growth plate, usually not completely through bone

Spiral fractures are indicative of abuse- often seen in the elderly or pediatric populations

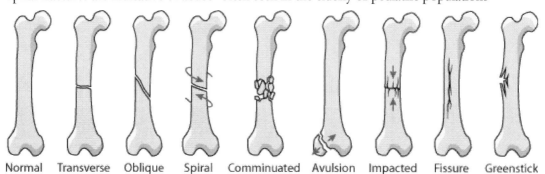

Normal   Transverse   Oblique   Spiral   Comminuated   Avulsion   Impacted   Fissure   Greenstick

(picture credit: http://askabiologist.asu.edu/how-bone-breaks)

Hydromorphone	Opioid Highly addictive	S/E: Sedation, hypotension, urine retention, constipation
Oxycodone	Opioid Highly addictive	S/E: Lightheadedness, sedation, constipation Admin with milk after meals
Hydrocodone	Opioid Highly addictive	S/E: Sedation, hypotension, constipation
Oxycontin	Opioid Highly addictive	S/E: Sedation, hypotension, constipation
Naloxone	Opioid reversal, IV, available by prescription nasal spray	S/E: sweating, nausea, vomiting, restlessness, headache

2. Fat embolism- after fracture of long bones (femur) when fat globules move into bloodstream, occurs within 24-48 hours

prevent by immobilizing fractures immediately

s/s: shortness of breath, chest pain, anxiety, decreased oxygen saturation, petechiae

tx: turn pt on left side in Trendelenburg, give O2 10L by mask, call physician

3. Compartment Syndrome

def: due to a crush injury, the muscle swells and is contained by the fascia, cutting off circulation

s/s: pain, paresthesias, pallor, pulselessness, pressure in extremity – requires a fasciotomy

(picture credit: http://wikipedia.org)

4. Traction

Purpose: immobilizes, alleviates muscle spasms, prevents deformities

Nrsg: Maintain straight alignment, weights hang free AT ALL TIMES, clean pins with peroxide and sterile swabs, bed blocks are okay to prop feet up, do not elevate knee gatch on bed

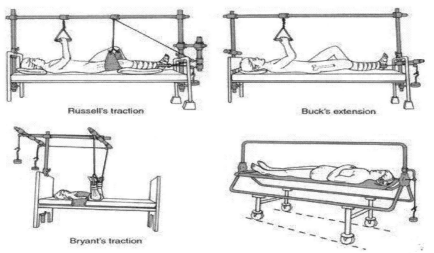

(picture credit: http://medical-dictionary.thefreedictionary.com/traction)

5. Cast Care

-avoid covering until 48 hours after application and elevate above heart until dry

-handle with palms while still wet

-check pulses- cap refill should be < 3 sec

-do not scratch skin under cast, report foul odor

-"hot spots" under the cast may indicate infection

-isometric exercises recommended- promotes venous return and circulation

6. Ambulatory Equipment

Crutches

    a.   4-point gait
        * Safest gait
        * **Requires weight bearing on both legs**
        * Move RIGHT crutch ahead (6 inches)
        * Move LEFT foot forward at the level of the RIGHT crutch
        * Move the LEFT crutch forward
        * Move the RIGHT foot forward

b. 3-point gait
* **Requires weight bearing on the UNAFFECTED leg**
* Move BOTH crutches and the WEAKER LEG forward
* Move the STRONGER leg forward
c. 2-point gait
* Faster than 4-point and requires more balance
* **Partial bearing on BOTH legs**
* Move the LEFT crutch and RIGHT foot FORWARD together
* Move the RIGHT crutch and LEFT foot forward together

-stairs: Up with the good, down with the bad

-measure two inches below axilla to measure crutch height

-support weight on hands rather than axilla

-correct angle of elbows: 20 to 30 degrees

-when moving, crutch should be 1 foot-length ahead of pt's feet

Cane

-handle must be level with hip bone

-cane should start 4-6" from little toe

-move cane and bad legs simultaneous

-hold cane in hand opposite the affected limb

Walker

-lift, then advance, weak leg, strong leg (LAWS)

-flex elbows at 20-30 degree angle when standing

-lift and move walker forward 8-10 inches

7. Hip Fracture/Replacement

-fractured leg will be shorter, externally rotated, adducted

-ice to site pre and post surgery

-encourage lifting onto bedpan

-log roll when moving

-do not sleep on operative side

-don't flex hip more than 45 to 60 degrees

-do not cross legs, use abductor pillow

-after 6 weeks, able to resume normal ADLs. Until then, no driving, no low toilet seats, no crossing legs

8. Amputation

Immediate post-op: observe for excessive oozing, elevate for first 48 hours, early ambulation, dressing changed daily until sutures are removed

stable care: wash, expose to air, turn patient to prone position frequently

Gabapentin	Indications: Seizure, RLS, nerve pain and/or neuropathy	S/E: Dizziness, drowsiness, aggression or trouble concentrating, fever, rash

9. Rheumatoid Arthritis v. Osteoarthritis

Rheumatoid Arthritis

s/s: systemic inflammation of joints- redness, symmetric joint pain and swelling, contractures, nodules

initially may begin as early morning stiffness

affects women more often than men

tests: ASO titer, C-reactive protein, sedimentation rate

nrsg: application of heat, anti-inflammatory drugs, use firm mattress, exercise as tolerated

Ibuprofen, Celecoxib	NSAIDs (anti-inflammatory) *Use with caution if pt has ASA allergy*	S/E: kidney toxicity in high doses -take with food
Adalimumab	Reduces inflammation (for RA or psoriatic)	S/E: Infections, numbness or tingling, weakness, problems with vision, decreased WBCs/platelets, fever that does not go away, bruising

Osteoarthritis

s/s: nonsystemic, contractures, stiffness, obesity, stress on joints

tests: MRI showing narrowing of joint spaces

Tx: corticosteroid injected locally, avoid flexion

10. Gout

s/s: swollen, reddened painful joints often in the great toe

limitation of motion

tests: xray, uric acid level

-eliminate purine food from diet- liver, sardines, seafood, meats, whole grains, alcohol

Allopurinol	Decreases levels of uric acid in blood, chronic treatment	S/E: Skin rash, joint/muscle pain, nausea adverse effect: Stevens-Johnson Syndrome
Colchicine	Decreases levels of uric acid in blood, acute treatment	S/E: Diarrhea, nausea, cramping, abdominal pains

11. Paget's Disease

def: a disease where new, spongy bone gradually replaces the original well-formed bone, usually localized to a few specific bones

causes: viral, genetic, environmental

s/s: deformities, bones that fracture and then re-heal incorrectly, pain in affected area

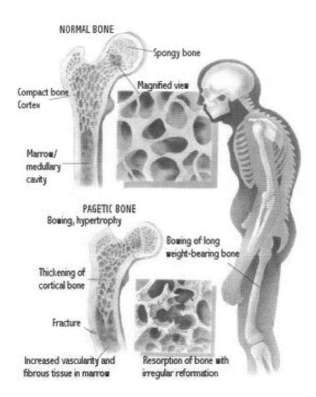

12. Osteoporosis

s/s: reduction in amount of bone mass

risk factors: small-framed, Asian, menopause, vitamin D deficiency, immobility, smoking

test: xray, bone density test

> Bone Density Test/Dexa Scan- remove all metal, okay to eat/drink

nrsg: provide optimal nutrition, encourage weight-bearing exercises on long bones (walking, NOT swimming), calcium supplement

Alendronic acid	Bisphosphonate Take meds in AM with water, remain upright for 30 minutes	S/E: Arthralgia, esophagitis
Calcium carbonate	Calcium supplement	S/E: Dysrhythmias, constipation, monitor EKG

Musculoskeletal Review Quiz

1. How are you as a nurse able to prevent fat emboli?

2. 5 Ps of compartment syndrome?

3. Most important thing to remember with traction?

4. How long should a cast be elevated?

5. When is a patient who had a hip fracture able to resume normal ADLs?

6. Specialty position for amputations?

7. Is rheumatoid arthritis isolated to one joint or is it systemic?

8. Osteoarthritis will show what on the x-ray/MRI?

9. What joint does gout begin in?

10. Encourage what type of exercise for osteoporosis?

11. Describe how to correctly ambulate with a two-point crutches gait.

12. What is the only crutches gait that may be used for a limb that is COMPLETE non weight bearing?

<u>Respiratory</u>

1. Atelectasis

- results from bad airway clearance, ie: sitting in bed without deep breathing

-pts most at risk are those bedridden after surgery, those in pain who do not want to deep breathe, especially due to abdominal surgeries

2. Use of accessory muscles

Accessory muscles = neck, shoulder, and clavicular area muscles to help breathing

-indicates extreme work of breathing and dyspnea

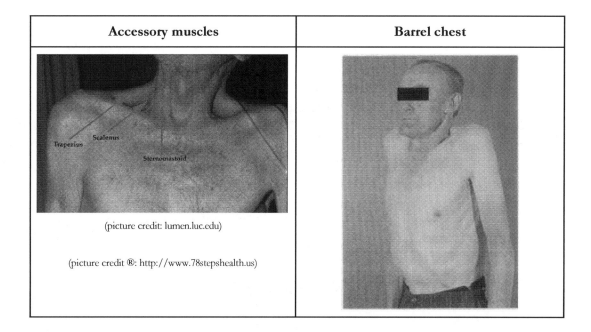

Accessory muscles	Barrel chest
(picture credit: lumen.luc.edu) (picture credit ®: http://www.78stepshealth.us)	

3. Rates of Breathing

Hyperventilation- low CO_2 -> have patient breathe into a paper bag

Hypoventilation – high CO_2 -> have patient increase depth and rate of respirations

4. Orthopnea - inability to breathe unless upright

-common in heart failure

-might ask: "how many pillows?"

5. Respiratory Sounds

Sound	Description	Patient Condition
Crackles/Rales	Gurgling, bubbling	HF, pneumonia, respiratory infections, fluid overload
Wheezes	High-pitched, musical, expiration	Asthma, anaphylaxis
Rhonchi	Low-pitched, coarse, snoring	Post-ictal phase of seizure, COPD, cystic fibrosis
Stridor	High-pitched, harsh, inspiration	Object in airway, anaphylaxis
Pleural Friction Rub	Grating sound on inspiration	Infection of pleural sac

6. Suctioning

-hyperoxygenate for 3 minutes before and after with deep breaths

-elevate HOB to semi-Fowler's position

-lubricate catheter with sterile saline, insert WITHOUT applying suction

-withdraw catheter and intermittently apply suction for no more than 10 seconds

-suction NOSE before MOUTH

7. Tracheostomy

-do not remove old ties before new ties are in place- change PRN

-sterile procedure to clean inner cannula q8hr

-apply sterile dressing over site- change q8hr and PRN

8. Mechanical Ventilation

-sedation is good to relax patients and ensure compliance

-check for hypoxia signs- restlessness, anxiety, increased HR

-most minimal vent setting: synchronized intermittent mandatory ventilation (SIMV)

-when alarm goes off for low pressure or high pressure, check the tubes before moving on to other options

    *HOLD - high pressure, obstruction, low pressure, disconnection

nsg: elevate head 30 degrees, assess readiness for extubation daily, oral care with chlorhexidine a minimum of daily → to decrease VAP

9. COPD, Emphysema – overinflation of alveoli, retained air

s/s: use of accessory muscles, weight loss, dyspnea, cough, clubbed fingers, prolonged expiratory phase, abnormal ABGs

Tx: administer LOW-FLOW oxygen- only 2L to prevent CO_2 receptor narcosis, encourage fluids, pursed lip breathing technique, incentive spirometer

Dextromethorphan	Antitussive Monitor cough type and frequency	S/E: Drowsiness, dizziness
Guaifenesin	Expectorant Take with a FULL glass of water	S/E: Dizziness, headache

10. Asthma – mucus and narrowed airways d/t allergic or stress stimulus

s/s: SOB, wheezing

Tx: administer Albuterol, sit up straight in bed

Cx: if there are decreased breath sounds or absence of wheezing, EMERGENCY!

Cx: Status Asthmaticus- acute exacerbation of asthma that remains unresponsive to tx

     -tx with theophylline, inhaled beta-adrenergics, IV corticosteroids, supp O2

Ipratropium	Long term tx of bronchospasm, neb	S/E: dry mouth, tachycardia
Albuterol	Short term tx of bronchospasm, neb	S/E: nervousness, hyperactivity, tachycardia
Aminophylline	Short term tx of bronchospasm , IV, based on weight of patient- can become toxic	S/E: nervousness, dizziness, tachycardia

11. Pneumothorax

-can be spontaneous: typical patient is young, tall, skinny male

s/s: acute onset of chest pain and SOB

Tension- shifts heart and great vessels- look for deviation of trachea toward unaffected side, jugular vein distention

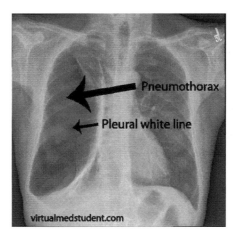

12. Chest Tubes

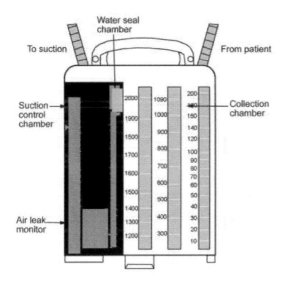

(picture credit:sinaiem.com)

-gentle intermittent bubbling in water seal chamber = air is gently leaving the patient

-vigorous and/or continuous bubbling in water seal chamber = air leak

-no bubbling in water seal chamber = pt's lung has re-expanded or the tube is occluded

suction control chamber = should remain constant and is PER MD order

- chest tubes are only clamped momentarily to check for air leaks, NEVER milked

- when physician removes chest tubes, nurse should encourage patient to do valsalva maneuver (bear down and hold breath)

-Cx: if tube comes out of patient, slap on occlusive dressing and tape all but one side

-Cx: if tube dislodges from collection chamber, cut off dirty end and put it in a container of sterile water

13. Pulmonary Embolism- blockage (may be made of blood clot, air, or fat) that lodges in the pulmonary artery

s/s: sudden SOB, chest pain, tachycardia, tachypnea, hemoptysis, low-grade fever

tx: Left side, trendelenburg, O2, call MD, anticoagulation

14. Pulmonary Edema- fluid buildup in the lungs, usually result of HF

s/s: restlessness, SOB, coughing up "pink tinged foamy sputum"

tx: semi-Fowler's position, diuretics (see Heart Failure)

15. Pleural effusion- fluid buildup around the lungs, HF most common

s/s: chest pain on inspiration, SOB, non-productive cough, crackles

tx: thoracentesis

16. Thoracentesis- procedure to remove excess fluid from between the lung and the chest wall

-procedure is done at bedside, verify informed consent, sterile technique

-position: pt sits on side of the bed, leaning over a bedside table with a pillow under arms

-physician should NOT drain effusion completely- increases likelihood of pneumothorax

   *no more than 1,000mL should be removed at one time

-when draining is complete, have pt take a deep breath and hum as needle is removed

17. Tuberculosis

s/s: night sweats, low-grade fever, cough with sputum

AIRBORNE PRECAUTIONS- N95 mask

-negative pressure room with 6-12 air exchanges per hr

-6 or 9 month regimen of 2-3 of the drugs below

dx: three negative morning sputums indicate TB is no longer active

-handwashing, put used tissues in a plastic bag

Isoniazid (INH)	Anti-tuberculosis Check LFTs, avoid alcohol, do not combine with Dilantin	S/E: hepatitis, peripheral neuritis, fever *Give V B6 with it to prevent neuritis
Rifampin	Anti-tuberculosis Check LFTs	S/E: hepatitis, fever, orange urine, tears, saliva
Ethambutol	Anti-tuberculosis Check LFTs	S/E: hepatitis, appetite loss, dizziness, GI upset

PPD- (aka Mantoux test) given at 10-15 degree angle on ventral surface of forearm with no aspiration, and read between 48-72 hrs

-okay if pregnant

-PPD is positive in people with a good immune system if >10mm

-PPD is positive in people with a BAD immune system if >4mm

-PPD cannot be used to test for TB in someone from a foreign country who has received a "shot" for TB

18. Legionnaires' disease- pneumonia caused by legionella, a bacteria that likes to grow in warm stagnant water (eg. Hot tubs, hot water tanks in buildings, decorative fountains)

-spread by inhaling the bacteria, especially with a weakened immune system

s/s: fever, headache, muscle pain, cough, SOB, chest pain, GI symptoms, confusion

Tx: antibiotics

19. ABGs

pH (acid base balance) = 7.35 to 7.45
CO_2 (carbon dioxide) = alkalotic < 35 to 45 > acidotic
HcO_3 (bicarbonate) = acidotic < 22 to 26 > alkalotic

Is the cause compensated or uncompensated?

      1. Uncompensated: if the pH is anywhere outside the normal ranges (greater than 7.45 or less than 7.35)

      2. Compensated: pH is anywhere inside the normal ranges (Anything between 7.35 to 7.45)

Please identify the following and clarify if they are compensated or uncompensated.

pH = 7.33
CO_2 = 50
HcO_3 = 25

pH = 7.45
CO_2 = 30
HcO_3 = 21

pH = 7.48
CO_2 = 40
HcO_3 = 30

Respiratory Review Quiz

1. Why does atelectasis happen?

2. Hypoventilation has ___ CO_2.

3. Crackles are described as sounding like...

4. Stridor occurs on _____.

5. A patient can be suctioned for no longer than _____ seconds.

6. Rule with tracheostomy ties?

7. Signs of hypoxia in a patient on a ventilator?

8. In what patient is it normal to have abnormal ABGs?

9. Three main symptoms of tuberculosis?

10. Who is the typical patient for a spontaneous pneumothorax?

11. What do you do if a chest tube becomes unhooked from the drainage portion?

12. What do you do if a chest tube pulls out of the patient?

13. A patient with tuberculosis is on what precaution?

14. What indicates that tuberculosis is no longer active in a patient?

15. If the pH is acidic and the HC03 is acidotic, the patient is said to be in what state? Compensated or uncompensated?

Endocrine

1. Diabetes

Type 1- acute onset: before age 30

s/s: polyphagia, polyuria, polydipsia

-insulin-producing cells are completely destroyed

-requires insulin injection, NO oral meds

-Diabetic Ketoacidosis (DKA)- very high levels of ketones when the body has run out of insulin and begins burning fat for energy- thirst, frequent urination, high blood glucose, HIGH levels of ketones in urine, fruity odor to breath, Kussmaul's respirations, vomiting, confusion

tx: insulin drip (cx: infusing too fast -> cerebral edema), Kcl (because insulin causes K to move back into cells and can cause serum hypokalemia), hydration

Type 2- gradual onset: after age 30, obese

-decreased sensitivity to insulin

-treated with diet, exercise, hypoglycemic meds & insulin injection

-Hyperosmolar Hyperglycemic Nonketotic Syndrome (HHNKS)- very high blood sugar levels, dehydration, warm dry skin, extreme thirst, eventually seizures and death

*NO ketones present because there is some insulin present, just less sensitivity

tx: insulin drip, Kcl, fluids (see above)

Hypoglycemia- s/s- irritable, confused, cold, clammy, sweaty

Hyperglycemia- s/s- hot dry skin, headache, fruity odor to breath, stupor/coma

HbA1c- blood sample taken without fasting that measures diabetic control over past 3 months

-normal is 4-6%, anything greater is indicative of bad diabetic control

Nrsg recommendations:

-no bare feet, shoes with natural fibers, cut nails straight across

-pts need to rotate sites after injecting insulin to prevent lipodystrophy

-best site to administer insulin for max absorption is abdomen

-when mixing insulin, draw up clear (regular) then cloudy (NPH)

-needs yearly visits to ophthalmologist and podiatrist

-no heating pads or hot baths due to peripheral neuropathy

-when pt is sick, need for insulin INCREASES

-exercise DECREASES blood glucose, advise patient to eat prior and reduce insulin

-48 hours before surgery: stop Glipizide, Sitagliptin, Glyburide, Repaglinide

-morn of surgery: hold Glimepiride, Glucophage, Glipizide, Glyburide

Insulin lispro	Rapid-acting insulin	Onset: 5-15 min, peak: 1 hr, duration: 3 hr, only used in insulin pumps
Insulin R (regular)	Short-acting insulin	Onset: 30-60 min, peak: 2-3 hr, duration: 4-6 hr, this is the only insulin that can be given IV
NPH insulin	Intermediate-acting insulin	Onset: 2-4 hr, peak: 6-12 hr, duration: 16-20 hr
Insulin glargine (Lantus)	Long-acting insulin	Onset: 1 hr, peak: none, duration: 24 hr, do not mix with any other insulin
Glucophage	Oral hypoglycemic	S/E: Diarrhea, hypoglycemia, nausea, vom
Glucagon	Oral hyperglycemic	S/E: Nausea, vomiting, bronchospasm, may repeat q15 minutes

2. Oral glucose tolerance test

Version A:

-pt fasts overnight and then office gives pt measured amount of glucose

-lab tech draws blood samples drawn at 1-, 2-, 3-h intervals

-if 2 hr glucose >200mg/dL on 2 occasions → diabetes

Version B:

-pt fasts overnight

-lab tech draws blood sample

-if fasting glucose > 126mg/dL on 2 occasions → diabetes

3. Where is everything??

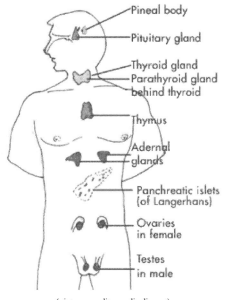

(picture credit: medindia.net)

4. Growth Hormone Problems (Pituitary Gland)

+ **Acromegaly**- s/s- lethargy, arthralgias, deep voice and enlarged tongue, enlarged flat bones, emotional instability, sexual abnormalities, hypertension, heart failure, hyperglycemia

Dx: growth hormone measured in blood plasma, MRI

Nrsg: Monitor blood sugar level, watch for signs of complications

surgery- hypophysectomy-

- mustache dressing under nose
- monitor for halo sign- clear drainage surrounded by yellow
- monitor for frequent swallowing- indicates bleeding
- instruct pt to avoid coughing/sneezing/blowing nose/bending forward, no brushing teeth for 2 weeks
- admin hormonal replacement as needed

- **Dwarfism**- s/s: height is below normal, proportions are norm, sexual maturity delayed, features delicate

Dx: Growth hormone measured in blood plasma

Nrsg: Monitor hormone replacement therapy and psychosocial

Human Growth Hormone (HGH)	IM, growth hormone	S/E: Headache, fluid retention, muscle aches -given daily at night time

5. Antidiuretic Hormone Problems (Pituitary Gland) (posterior)

+**SIADH**- s/s: anorexia, change in LOC, tachycardia, decr urine output

Dx: Small cell carcinoma of the lung, ADH high, high urine specific gravity, low serum sodium

Nrsg: Restrict fluid intake, admin diuretics, hypertonic saline (requires 2 RNs to cosign), monitor serum Na+ levels

-**Diabetes Insipidus**- s/s: excessive urine output, chronic dehydration, excessive thirst, weight loss, weakness, constipation

Dx: Deficiency of ADH, low urine specific gravity, high serum sodium

Nrsg: Monitor I&Os, administer Pitressin, DDAVP

Vasopressin	"Pitressin", ADH, given IV	S/E: Stomach pain, bloating, gas, dizziness
DDAVP	"Desmopressin", ADH, nasal spray	S/E: Confusion, nosebleeds, nausea, rapid weight gain, headache

6. Cortisol Problems (Adrenal Cortex)

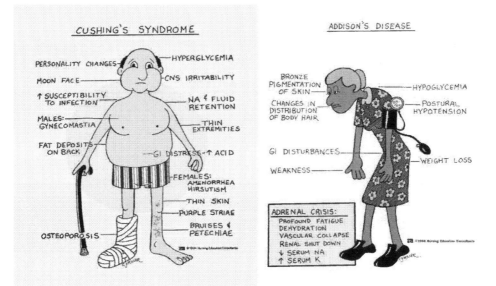

(picture credit: allfornursing.blogspot.com)

+ Cushing's disease- s/s: muscle wasting, osteoporosis, edema, purple striations on skin, truncal obesity, mood swings, blood sugar imbalance, hypertension, buffalo hump

<u>Dx</u>: Hypokalemia and hypernatremia, MRI

<u>Nrsg</u>: Teach appropriate diet- high protein, low carb, low Na, high K+, fluid restriction

<u>Tx</u>: Post-adrenalectomy care- check glucose levels, provide long-term hormone therapy

- Addison's Disease- s/s: fatigue and weakness, alopecia, tan skin, weight loss, fluid and electrolyte imbalance, dehydration, depression, hypoglycemia

<u>Dx</u>: Hyperkalemia and hyponatremia, MRI

<u>Nrsg</u>: Teach appropriate diet- high protein, high carb, high Na, low K+, avoid stressors

<u>Tx</u>: Admin hydrocortisone, fluids, check blood sugar, daily weights

Hydrocortisone	IM, corticosteroid	S/E: Truncal obesity, osteoporosis, mood swings, purple striations on skin, edema (moon face), hyperglycemia

7. Pheochromocytoma (Adrenal Medulla)

<u>def</u>: hypersecretion of catecholamines (epinephrine & norepinephrine) from tumors of the adrenal medulla

<u>s/s</u>: intermittent hypertension lasting sev mins to hrs, incr HR, palpitations, nau and vom, tremors and nervousness, pounding headache

<u>Dx</u>: Histamine test- normally causes drop in BP, but will in this case cause a rise

<u>Nrsg</u>:, administer phentolamine, avoid stress, analgesics, sedatives/tranquilizers for rest, incr calories in meals, avoid coffee/tea/caffeine, pre patient for adrenalectomy or medullectomy

Phentolamine	IV, alpha adrenergic antagonist	S/E: hypotension, flushing

8. Thyroid Problems (Thyroid Gland)

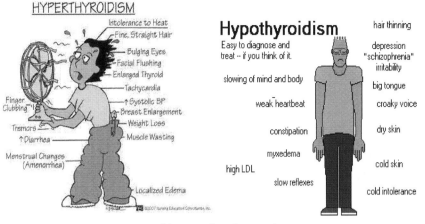

(picture credit: pathguy.com)

+Hyperthyroidism (Grave's Disease)- s/s: incr physical activity, tachycardia, incr sensitivity to heat, fine soft hair, nervous jittery, exophthalmos, weight loss, finger-clubbing

Tx: Antithyroid drugs, irradiation, thyroidectomy

Nrsg: Limit activities to quiet ones, provide rest, keep room cool, avoid stimulants (coffee)

Pre-thyroidectomy- may be treated with potassium iodide (KI) to reduce the vascularity of the thyroid (reducing bleeding)- will take effect in 1-2 weeks

Post-thyroidectomy- prevent strain on suture line with, support head and neck, keep trach set and suction supplies at bedside, side effect of hoarse, weak voice is normal

Cx: thyroid storm- high temp, high BP, tachycardia, exacerbated symptoms, can lead to death

-Hypothyroid- s/s: cold intolerance, lethargy, tiredness, constipation, weight gain

*can be caused by autoimmune disorder- known as Hashimoto's thyroiditis

Nrsg: allow pt extra time for activities, frequent rest periods, maintain warmer room temp, promote comfort, rest, sleep, high protein and low cal meals, no soap, use creams

Tx: Synthroid (artificial TH), admin sedatives cautiously, warn client to never d/c Synthroid abruptly or could turn into myxedema coma- hypothermia, low blood pressure, bradycardia, exacerbated symptoms, death

Synthroid (Levothyroxine)	Thyroid Replacement- Take in AM, monitor weight, get baseline vitals	S/E: Nervousness, palpitations, insomnia
Propylthiouracil (PTU)	Thyroid Suppressant	S/E: Leukopenia, fever, jaundice

9. Parathyroid Problems (Parathyroid Gland)

+Hyperparathyroid- s/s: fatigue, muscle weakness, cardiac dysrhythmias, renal calculi, pathological fractures, pancreatitis, peptic ulcer

Dx: elevated serum calcium, decr serum phosphorous, xray- bones appear porous

Nrsg: increase fluid intake to decr risk of renal calculi formation, prevent fractures, monitor potassium levels (which counteracts eff of calcium on cardiac muscle), provide post-parathyroidectomy care, IV Lasix to promote calcium excretion

-Hypoparathyroid- s/s: tetany, muscular spasms, dysphagia, paresthesia, anxiety, depression, tachycardia, + Chvostek's and + Trousseau's sign

       Chvostek's sign- spasm when tapping face- low calcium
       Trousseau's sign- flexion of wrist when BP cuff inflated- low calcium

Dx: Decr serum calcium, incr serum phosphorous, xray- bones appear dense

Nrsg: emergency tx- calcium gluconate over 10-14 mins, observe for tetany

Endocrine Review Quiz

1. Type one diabetes has onset... when?

2. Oral hypoglycemics cannot be given to what type of diabetes patient?

3. Signs and symptoms of hypoglycemia? Hyperglycemia?

4. Normal HbA1C?

5. Blood samples for GTT are drawn at what intervals?

6. SIADH is associated with what cancer?

7. Signs and symptoms of diabetes insipidus?

8. Hypertonic saline is given to a patient with...

9. The histamine test diagnoses what?

10. Clinical labs of a patient with Cushings?

11. Appropriate diet for a patient with Addison's?

12. A patient with Grave's disease will receive what?

13. What is Chvostek's Sign? Trousseau's?

14. What time of day is Synthroid given?

15. A patient with hypoparathyroid will most seriously have what complication?

Neurology

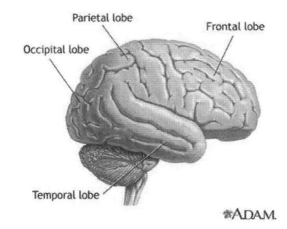

Parietal lobe
Occipital lobe
Frontal lobe
Temporal lobe

*A.D.A.M.

1. Physical Components of Brain

Right/left hemispheres = each controls opposite sides

Frontal lobes- Broca's area- formation of words/speech

Prefrontal/frontal areas- personality

      s/s of injury: changes of personality, memory impairment, poor concentration

Parietal lobe- concept of body image and awareness of external environment

      s/s of injury: numbness, sensation confusion, right-left disorientation, apraxia, trouble

      drawing/constructing things, easily get lost in their own neighborhood

Occipital lobe- visual center

      s/s of injury: blindness, difficulty recognizing faces, confabulation

Temporal lobe- dominant hearing of language, Wernicke's area- recognition of language

      s/s of injury: difficulty comprehending language, diminished hearing, amnesia of recent

      memories/events

Hypothalamus- responsible for temperature, blood pressure, sleep, appetite

Cerebellum – fine and gross motor skills, balance

Limbic systems- emotional behavior

Brain stem- respiratory and vasomotor activity

2. Cranial Nerve Assessment

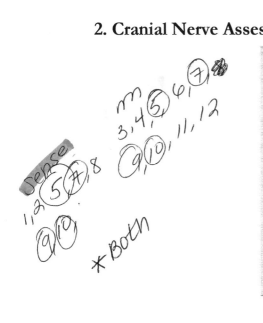

(picture credit: americannursetoday.com)

Oh, **O**h, **O**h, **T**o **T**ouch **A**nd **F**eel **V**irgin **G**irls **V**aginas, **S**uch **H**eaven!

I. Olfactory- sense of smell
II. Optic- sense of vision
III. Oculomotor- symmetry of eye opening, pupil dilation
IV. Trochlear – eye motor movements
* V. Trigeminal – jaw movement and sensation _→ eye move 6 muscle_
VI. Abducens – eye motor movements
* VII. Facial – facial sensory and movement _taste/facial express_
VIII. Vestibulocochlear (Acoustic) – hearing intact, balance
* IX. Glossopharyngeal – gag reflex, ability to swallow, ability to taste
* X. Vagus – Able to swallow and speak with a smooth voice
XI. Spinal accessory – flexion of head, shrugging of shoulders
XII. Hypoglossal – motor tongue movements

3. Head Injury

<u>s/s:</u> Battle's sign(ecchymosis over mastoid bone artery), raccoon eyes, rhinorrhea, otorrhea, possible change in level of consciousness, contusion

<u>Tx:</u> management of increased ICP, hypothermia, admin glucocorticoids, barbiturate therapy (see Seizures)

4. Increased Intracranial Pressure

<u>s/s:</u> cerebral edema, **altered level of consciousness (first sign!),** observe seizure precautions, elevate head 15-30 degrees, prevent valsalva maneuver, exhale when moving in bed, monitor vital signs hourly- be alert for widening pulse pressure

<u>nrsg:</u> restrict fluids 1200-1500 ml/day, do not restrain, provide environment that will prevent injury

Mannitol	Diuretic specific to brain/head area	S/E: Dizziness, hypotension, dehydration
		nsg: drug crystallizes at room temp, use filter needle

5. Seizures- abnormal electrical activity in the brain

-can occur for a number of reasons and be related to or independent of comorbidities

-may be petit mal (no motor movement, "absence") or gran mal (convulsions present)

-have a pre-ictal, ictal, post-ictal phase

dx: EEG-

>pre: explain procedure, painless, avoid stimulants for 24 hr prior, may be kept awake the night prior

>post: help client wash hair, observe for seizure activity

nsg: keep bed lowered to floor, place mattress/pads around bed or padded side rails

-during a seizure: no tongue blade, lower patient to floor, pad head, remove restrictive clothing, administer oxygen if needed, do not leave the patient

-after a seizure: suction mouth if necessary, keep client covered and warm, take vital signs, document everything, assess for complications

tx: anticonvulsants

**All anticonvulsant, antidepressant, and antipsychotic medications have the potential to lower the CBC and reduce WBCs, RBCs, and platelets. **

Carbamazepine	Anticonvulsant	S/E: Myelosuppression, diplopia, monitor CBC, Stevens-Johnson
Clonazepam	Anticonvulsant, benzodiazepine	S/E: Confusion, drowsiness
Topiramate	Anticonvulsant	S/E: Vision problems, renal calculi
Trileptal	Anticonvulsant	S/E: decreases birth control effectiveness
Diazepam	Anticonvulsant, benzodiazepine (Valium)	S/E: Hypotension, do not take with alcohol
Levetiracetam	Anticonvulsant	S/E: Dizziness, +SI
Phenytoin	Anticonvulsant (Dilantin) Level: 10 - 20	S/E: Hirsutism, GI upset, gingival hypertrophy, brown/red sweat/urine, depletes vit D, skin rash, ataxia, slurred speech, contraindicated in pregnancy
Phenobarbital	Anticonvulsant	S/E: Drowsiness, constricts pupils, contraindicated in pregnancy
Valproic Acid	Anticonvulsant	S/E: Sedation, tremors, prolonged bleeding time, monitor platelets, bleeding time, liver functions

6. Glasgow Coma Scale

Glasgow Coma Score		
Eye Opening (E)	**Verbal Response (V)**	**Motor Response (M)**
4=Spontaneous 3=To voice 2=To pain 1=None	5=Normal conversation 4=Disoriented conversation 3=Words, but not coherent 2=No words......only sounds 1=None	6=Normal 5=Localizes to pain 4=Withdraws to pain 3=Decorticate posture 2=Decerebrate 1=None
		Total = E+V+M

(picture credit: ophthalmology.stanford.edu)

-best score: 15, worst score 3, <8 = coma

-<u>decorticate</u> – everything pulls into the core and flexed (toward the in)

    -sign of damage to the nerve pathway between brain and spinal cord that could heal

-<u>decerebrate</u>- everything fully extended (toward the out)

    -sign of damage to brainstem, most likely unable to heal

7. General Imaging

<u>CT Scan</u>

Pre: written consent, explain procedure, immobile during exam, if contrast dye used, may experience flushing, warm face and this is normal

Post: encourage PO fluids

<u>MRI</u>- pre: screen pt for pacemaker or metal

<u>Myelogram</u> - xray of spinal cord/canal after injection

Pre: informed consent, explain procedure, NPO for 4-6hr before test, obtain allergy history

Post: neurologic assessment every 2-4 hr, oral analgesics for headache

8. Trigeminal Neuralgia- def: neuralgia involving the fifth cranial nerve

<u>s/s</u>: stabbing, burning facial pain, twitching, grimacing of facial muscles

<u>nrsg</u>: avoid stimuli (light touch) that exacerbates attacks, admin pain relievers

<u>tx</u>: Gabapentin (see Musculoskeletal)

9. Bell's palsy- def: facial paralysis involving the seventh cranial nerve

<u>s/s</u>: inability of eye to close, incr lacrimation, speech difficulty, loss of taste, distortion of one side of face

<u>nrsg</u>: protect head from cold or drafts, improve facial muscle tone with electric stimulation, prevent corneal abrasions, heat okay

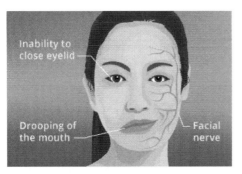

(picture credit: allaboutvision.com)

10. Acoustic neuroma - def: benign tumor of the eighth cranial nerve

s/s: deafness (partial initially), tinnitus, dizziness

tx: surgical excision of tumor

11. Spinal Cord Injury

s/s: flaccid paralysis of skeletal muscles, postural hypotension, bradycardia, edema, alterations in normal temp regulation

C1-C4 at increased airway/breathing risk!!! no bladder control

T1-T5 – paraplegia, no bladder control, maybe weakened arms

L1-L5- weakened legs, no bladder control

S1-S5- most likely able to walk, no bladder control

nrsg: ensure patent airway, move client by log-rolling technique, steroid therapy, bladder and bowel training

Cx: Autonomic dysreflexia- SWEATING, headache, bradycardia, hypertension, flushing above level of injury, nasal stuffiness, anxiety

Tx: 1. IMMEDIATELY SIT PT STRAIGHT UP IN BED

2. CHECK FOLEY CATHETER OR CATHETERIZE, POSSIBLY DISIMPACTION

3. ADMINISTER ANTI-HTN

11. Herpes zoster (Shingles)

def: the chicken pox virus (herpes zoster) reactivates in an adult and affects a specific nerve area

s/s: severe pain in specific dermatome for 1-2 days before skin changes, vesicular blisters, muscle weakness in the affected area, headache

tx: antiviral

Acyclovir	Antiviral	S/E: Headache, seizures, diarrhea

12. Meningococcal Meningitis

s/s: headache, fever, photophobia, nuchal rigidity, + kernig's sign, + brudzinski's sign, alterations in mental status, seizures, increased ICP

dx: lumbar puncture, unless ICP increased (normal level is 5-15)

nrsg: provide nonstimulating dark environment, seizure precautions, droplet precautions

Lumbar puncture

Pre: explain procedure, obtain informed consent, bedside procedure, position in lateral recumbent fetal position

Contraindication: increased ICP, in which case should not do LP

Post: Neuro assessment, position flat 4-12hr, encourage PO fluid to 3,000mL, oral analgesics for headache

13. Alzheimer's Disease/Dementia

s/s: loss of memory, spatial judgment, visuospatial perception, and personality

stages:

a. Early- Forgets names, inability to make decisions, unable to travel alone to new destinations

b. Moderate- Gross intellectual impairments, complete disorientation to time, place, events, loss of ability to care for self

c. Severe- Completely incapacitated, motor and verbal skills lost, totally dependent ADLs

nrsg: structure the environment, promote independence in ADL, promote bowel and bladder continence, encourage reminiscing

tx: medications for behavior (see antipsychotics) and Alzheimer's-specific (below)

Memantine	Treats dementia, memory loss	S/E: Hallucinations, confusion, fatigue
Donepezil	Treats dementia, memory loss	S/E: Nausea, bleeding, dizziness

Neurology Review Quiz

1. Broca's area of the brain is responsible for _____ while Wernicke's area is responsible for _____.

2. Where is personality located in the brain?

3. List and describe the twelve cranial nerves.

4. List three signs for increased ICP. Then list five interventions.

5. A patient has a headache after a lumbar puncture. Do what first?

6. Patients who are allergic to _____ should not receive contrast dye.

7. Bell's palsy involves what cranial nerve?

8. What is Battle's sign? What does it mean?

9. Signs of autonomic dysreflexia? Interventions?

10. Most important intervention in Guillain Barre?

Eyes, Ears, and Nose

1. Eyes

Snellen Test: a measure of vision via eye chart

        -numerator: distance that person stands from the chart (normally 20 feet)

        -denominator: distance a person can read a particular line

                *should be "20" but if a person can only read the giant E at 20 feet, which a

                normal person can read at 200 feet, their vision is now "20/200"

myopia- nearsightedness (can only see things up close, not far away)

presbyopia- farsightedness (cannot focus eyes on things up close), occurs with aging

Cataracts: lens of the eye becomes progressively opaque

        -s/s: gradual, painless blurring of central vision

        -post cataract extraction and intracapsular implantation:

                -severe pain/nausea may be a sign of increased ICP/hemorrhage

                -avoid vomiting or increasing head pressure, no vacuuming or golf, no lifting more than 5lbs, do not sleep on operative side

                -hair can be shampooed 1-2 days after surgery with head tilted back

                -mild itching is normal

Glaucoma: increased pressure in eye that causes damage

        -s/s: gradual loss of peripheral vision

        Note: vision loss is irreparable once it is present

        -tonometry- measures the pressure in the eye. Normal: 10-21

        -eye appt to measure tonometry should be in AM, when pressure is the highest

Atenolol 4% Timolol 4%	Eye drops to tx glaucoma	S/E: Lowered BP, reduced pulse, fatigue, depression
Latanoprost	Eye drop to tx glaucoma "Xalatan"	S/E: Stinging/burning eyes, chest pain, blurred vision

Macular Degeneration: macula of the eye becomes damaged, related to age

        -will lose a large amount of central vision, but keep peripheral

        -permanent and chronic vision loss, but may be treated to slow progress

Retinal Detachment: retina detaches from rest of eye, cause is usually unknown

        -s/s: feeling that a shadow or curtain is falling across the field of vision, "red floaters", abrupt flashing lights-ophthalmic emergency!

        Tx: cover only affected eye after surgery, limit movement of both eyes, no bending/coughing/sneezing, orient pt, give stool softener

Foreign Object:

        -do NOT remove in ER. Only ophthalmologist may remove

Chemical Burn:

        -irrigate immediately with sterile normal saline for at least 10 minutes

        -assess visual acuity

<u>Red light reflex</u>- caused by reflection of light off the inner retina, if absent, neuro problem

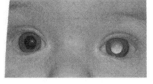

<u>Blind patients</u>- use imaginary clock face to describe food placement at meal times
-walk slightly ahead and to the side of a blind patient
-offer arm for the blind patient to take

2. Ears

<u>Otitis media</u> (children): ear infection, often progresses from a standard cold
        <u>s/s</u>: tugging at ear, crying, pain, discharge, fever
        <u>tx</u>: antibiotics

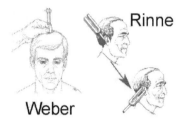

<u>Weber Test</u>:
        -purpose: to detect sensorineural hearing loss
        -to perform, place the base of the fork on the midline of the patient's skull. Hit
        fork. Pt is asked to report in which ear the sound is louder.
        -**normal:** SHOULD be about the same.

<u>Rinne test</u>:
        -purpose: to evaluate unilateral hearing loss
        -to perform, place a vibrating tuning fork against the patient's mastoid bone. Ask
        the patient to tell you when the sound is no longer heard. When signaled, quickly
        position the still vibrating tuning fork 1-2cm from the auditory canal and again ask
        if able to hear
        -**normal**: patient should be able to hear the tuning fork held next to the ear after
        they can no longer hear it when held against the mastoid bone

<u>Meniere's Disease</u>: disorder of the inner ear that affects hearing/balance
        <u>s/s</u>: episodes of vertigo, low-pitched tinnitus, hearing loss that fluctuates
        <u>Tx</u>: safety, low sodium diet, slow head movements, have pt listen to music, keep in
        a dark room

<u>Ear exams</u>: pull the pinna up and back in an adult, down and back in a child <3yr/old

<u>Deaf/Hearing-Impaired patients</u>:

-demonstrating or pantomiming may assist in communication

-ask for the patient's preferred way to communicate

-speak directly and slowly, not loudly

3. Nose

<u>Patency of nostrils</u>
-assessed by having patient sniff inward through one nair while the other is occluded

<u>Anterior epistaxis</u>: nosebleed
<u>Tx</u>: position upright and leaning forward to prevent blood from entering the stomach, apply direct lateral pressure to the nose for 5 minutes, apply ice or cool compresses to the nose, instruct patient not to blow nose for several hours

<u>Rhinoplasty</u>: nose surgery, for cosmetic or medical reasons (ie, fix a deviated septum)
-do not remove packing
-do not take oral temperature (because nose will be occluded)
-do not blow nose or increase pressure
-any packing can lead to Stevens-Johnson

Ears, Eyes and Nose Review Quiz

1. A nurse administers the Snellen test to a patient and reports results of 20/60. Translate what this means.

2. A patient with Meniere's disease should adhere to what type of diet?

3. You are the scheduler for an ophthalmologist's office. You receive a call from a patient who says they need to make an appointment for a tonometry screening. What appointment time is best?

4. Describe how to perform the Rinne Test.

5. What type of hearing loss does the Weber Test identify?

6. What is a normal tonometry reading?

7. What is the correct procedure for a nurse to remove a foreign body from a patient's eye?

8. What is the correct action for a patient with a chemical burn to the eye?

9. A child cries, tugs at his ear, and has a fever of 100.9 degrees. Diagnosis?

10. What is epistaxis?

Gastrointestinal

1. Diets

a. **Clear Liquid**
-tea, coffee, broth, popsicles (no milk or fruit juice with pulp)

b. **Full Liquid**
-clear liquids, custards, pudding, ice cream, creamed soup

c. **Bland Diet**
-milk, custards, cereals, white bread, creamed soup, potatoes
-NO highly seasoned foods, tea, coffee, citrus fruits, meat soups, raw fruits, veggies, whole grains

d. **Low Residue Diet**
-Clear fluids, sugar, salt, meats, fats, eggs, limited milk, cereals, white bread, peeled white potatoes
-NO cheeses, fried foods, highly seasoned foods, HIGH FIBER

e. **High Fiber Diet**
- raw fruits, veggies, whole grains, foods low in carbs

f. **Diverticulitis Diet**
-NO corn, nuts, seeds, whole grains

g. **Low Fat Diet**
- Skim milk, fruits, veggies, bread, cereals
- NO whole milk, fried foods, eggs, avocado, cheese, chocolate, nuts, peanut butter, creamed soups, highly seasoned foods

h. **Low Sodium Diet**
- Salt-free preparations
- NO smoked/salted meats/foods, processed foods, most frozen veggies, seasonings, chocolate, beets, celery, spinach

i. **High Potassium Diet**
- Fruit juices, cocoa, dark leafy greens, avocados, cantaloupe

j. **Low Purine Diet**
- Most vegetables, spinach, fruit juices, cereal, eggs, fat free milk, cottage cheese
- NO meats, fish, fowl, gravies, lentils, nuts, beans, oatmeal, whole wheat, alcohol

k. **Chronic Renal Failure- Protein Restriction**
-can eat white bread, cereal, most other carb-heavy foods
-complete proteins only in limited quantities (like 4oz chicken breast), avoid Na, avoid K+, avoid phosphorous, no salt substitutes

Magnesium Hydroxide/ Aluminum	Antacid	S/E: Laxative effect
Magnesium hydroxide	Antacid	When excessive- nausea, vomiting, diarrhea
Cimetidine	Anti-ulcer	S/E: Diarrhea, confusion -Dose at bedtime
Bismuth subsalicylate	Antidiarrheal (Pepto Bismul)	S/E: Darkening of stools and tongue, constipation, give 2 hr before and 3 hr after other meds
Docusate	Stool softener	S/E: abdominal cramps
Metamucil	Stool bulk-forming	S/E: Obstruction of GI tract *Take with FULL glass of water.
Diphenoxylate	Antidiarrheal, decr peristalsis	S/E: Sedation, tachycardia
Prochlorperazine	Antiemetic	S/E: Orthostatic hypotension, photosensitivity
Metoclopramide	Antiemetic	S/E: Restlessness, EPS symptoms, take b4 meals
Ondansetron	Antiemetic	S/E: Diarrhea, headache, dizziness

2. Nasogastric Tube

purpose: gastric decompression, relief of symptoms, prevents aspiration

-daily gastric pH can be used to verify placement- ph between 0-4

-cx: abdominal distention – check suction, reposition tube, then try to irrigate

-must have order to remove

-does not come out post-op until peristalsis present

True drain = total amt minus what you used for irrigation

3. Tube Feeding

-xray verifies tube placement

-if aspirate stomach >100mL residual, hold feeding

-replace aspirate after examining/measuring

-cx: diarrhea

True drain = total amt plus what you used for irrigation

-change bag/tubing q24hr

4. TPN

-used through central administration-> watch for SOB, pain, anxiety - > air embolism

tx: turn on left side in trendelenburg, give 10L O2 via mask, call physician

cx: INFECTION

nsg: accuchecks q6hr, weight, labs, dressing change are 2-3 x per week

-prevent change to rate infusion or else -> hyper/hypoglycemia

-to stop TPN, gradually decrease rate to prevent rebound hypoglycemia

-if specific TPN type is unavailable, give 10% dextrose IV fluids until TPN is available

-change bag/tubing q24hr

5. Hiatal Hernia

Def: upper part of stomach moves up through the diaphragm

s/s: feeling of fullness, epigastric pain, belching, heartburn

nrsg: do not lie down for at least an hour after meals, elevate head of bed when sleeping

6. GERD

Def: reflux of stomach acid into esophagus

s/s: belching, heartburn, burning in chest, nausea

nsg: avoid acidic foods, 5 small meals, elevate HOB 30 degrees

cx: Barrett's esophagus

7. Ulcers of GI Tract

cause: H. Pylori bacteria; ulcers can be located in stomach (gastric) or intestines (duodenal)

s/s: bloating, heartburn, nausea, pain in stomach

*pain immediately after eating → gastric ulcer

*pain a few hrs after eating→ duodenal

nsg: most foods are OK as tolerated

dx: Esophagogastroduodenoscopy (EGD)

8. Anatomy

Liver- upper right side

Gallbladder- right side, under liver

Pancreas- left side

Spleen – lower left side

Appendix- lower right side

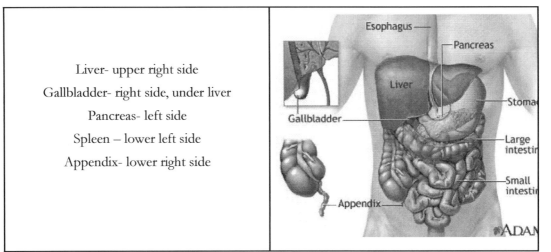

9. Cirrhosis

s/s: anorexia, nausea/vomiting, fatigue, yellow tint to skin, petechiae, dark urine, confusion

-can be caused by alcohol intake, drugs, Hep B or C, etc

tx: no cure, but we treat the various symptoms with palliative care

Cx #1: hepatic encephalopathy d/t ammonia levels→ lactulose, neomycin

Lactulose	Reduces ammonia through elimination	S/E: Diarrhea, loose stools, dehydration, dizziness
Neomycin	Reduces ammonia by killing bacteria in GI tract that produce it	S/E: Nausea, diarrhea, ototoxicity, nephrotoxicity

Cx #2: esophageal varices d/t portal hypertension → Sengstaken Blakemore tube

*keep scissors at bedside

Cx #3: ascites→ treatment is a paracentesis

Paracentesis

-performed at bedside with patient in semi-Fowler's position

-empty bladder prior to procedure

-afterward- check vital signs frequently in case of hypotension due to sudden fluid loss

Liver Biopsy

- performed at bedside, supine position, upper arms elevated

-admin of vitamin K

66

-NPO morning of test

-post: position on RIGHT side and flat for at least 4 hours

-maintain bedrest for 12-24 hours

-expect mild local pain and radiating to right shoulder

-report severe abdominal pain, board-like abdomen immediately

Vitamin K	Clotting agent	Phytonadione = Brand name
		S/E: DVT, clots, stroke

10. Cholecystitis

s/s: severe pain in RUQ that can radiate to back, intolerance to fatty foods, nausea, vomiting, flatulence, elevated temp, dark urine, clay-colored stools

nrsg: antibiotics, NPO until acute symptoms subside, gastric decompression, analgesics

tx: if gallbladder is removed, pt will have a T-tube where bile will drain from- at first, 500-1,000ml/day, may refer pain to R shoulder

Cholecystogram- checks gallbladder

-prep: fat-free dinner the evening before

-ingestion of dye tablet- check for allergies to shellfish

- NPO after dye ingestion, then xrays

-pt eats a high fat meal, then more xrays

Propantheline bromide	Decreases biliary spasms	S/E: dizziness, drowsiness, decrease sense of taste

11. Pancreatitis

s/s: nausea and vom, severe pain, fever, jaundice, hyperglycemia, weight loss, ascites

risk factors: alcohol, gallstones, abdom trauma, hyperlipidemia, elevated triglycerides

nrsg: NPO, gastric decompression, analgesics, antibiotics, fluids

+ Cullen's sign - ecchymosis in umbilical area, retroperitoneal bleeding
+ Turner's sign - ecchymosis of flank, retroperitoneal bleeding

12. GI Tests

a. Upper GI- Barium Swallow - pt swallows barium to check upper GI system with xrays

pre: maintain NPO after midnight

post: encourage fluids/laxatives to get out barium

b. Endoscopy- under sedation, camera on a wire is put down pt's throat to check for abnorms

-verify informed consent is signed

pre: maintain NPO before procedure for 8h

post: maintain NPO until gag reflex returns, inform pt to expect sore throat

c. Lower GI- Barium enema- pt receives barium enema and gets xrays

pre: low residue diet for 1-2 days as prep

-clear liq diet and laxative evening prior to test

-cleansing enemas until CLEAR, otherwise test cannot be done

post: cleansing enemas to remove barium, re-hydrate

d. Colonoscopy- under sedation, camera on a wire is inserted into pt's anus and up into GI tract

pre: pt drinks GoLitely until defecations are clear

-clear liq diet 12-24 hrs prior to exam, 6-8 hours of NPO

e. Decompression of the intestinal tract - like an ng tube, but lower

-requires Salem Sump or Miller-Abbott tube

f. Fecal Occult Blood Testing - stool sample to look for blood

pre: avoid red meat, poultry, fish, beets, broccoli, cauliflower, horseradish, mushrooms, turnips

-avoid iron, vitamin C, ASA, NSAIDs

-maintain a high-fiber diet to aid in "collecting specimens"

13. Irritable Bowel Syndrome (IBS)

-includes Crohn's and Ulcerative Colitis

s/s: diarrhea, abdominal cramping, blood in stools

nrsg: maintain hydration, high-protein and high-calorie, low residue diet

tx: steroids, TPN, bowel rest, antidiarrheal meds, no dairy

Adalimumab	Reduces inflammation	S/E: Infections, numbness or tingling, weakness, problems with vision, decreased WBCs/platelets, fever that does not go away, bruising

14. Diverticulitis

s/s: pain in LLQ, diarrhea, low grade fever, bowel irregularity

tx: NPO, antibiotics, IV, maintenance diet (see Diets section)

15. Intestinal Obstruction

s/s: high-pitched, scarce bowel sounds, abdom pain and distention, nau and vom

nrsg: NG or intestinal decompression, NPO, Fowler's position to facilitate breathing

16. Dumping Syndrome

Def: cx of stomach surgery (Billroth II procedure)

s/s: abdominal cramps, nausea, diarrhea

tx: eat small meals, do not drink fluids with meals, lie down after meals, increase consumption of foods with fiber/protein/fat and low carbs/salt

17. Colostomy and Ileostomy

-with an ileostomy, stool is liquid

-colostomy- formed to loose stools

preop: high calorie diet, NPO after midnight prior, laxatives

post: NG or intestinal decompression until peristalsis returns

f/u: -clear liq diet progressing to solid, low-res diet for 6-8 weeks

stoma: initially red, swollen, subsides to red/pink

-danger: blue, grey, dusky

-does not drain fecal matter for 3-6 days postop

irrigations: used for colostomy to encourage bowel training

18. Dehiscence and Evisceration

s/s: bowel and/or other organs protruding through the skin

-more likely to occur in overweight, diabetic clients who have had abdominal incisions

Tx: cover with moist sterile dressing and call physician

19. Appendicitis

s/s: abdom pain in right lower quadrant- McBurney's point, nau and vom, rigid abdomen and muscle guarding

nrsg: no heating pads, ice okay, maintain NPO, *sudden absence of pain indicates rupture

GI Review Quiz

1. Is coffee allowed on the bland diet?

2. A patient has just returned from an endoscopy and would like a drink of water. What do you tell them?

3. A patient scheduled to get a lower GI tract test is given a barium enema. 20 minutes prior to the test, they pass a brown, loose bowel movement. What do you do?

4. What is the absolute method to verify placement of a feeding tube?

5. Where on a patient would you palpate the liver? The pancreas?

6. List three dangers of changing the rate of TPN on a patient.

7. Explain the process of a cholecystogram. What organ does it focus on?

8. What side should a patient getting a liver biopsy be placed on afterward? What pain is normal and expected?

9. When is a T-tube used in a patient?

10. One of your patients has newly returned from surgery after having a bowel resection. They have a stoma protruding from their stomach that is a dusky blue color. What do you do?

11. What does McBurney's point refer to?

12. Name some differences between Crohn's disease and Ulcerative Colitis.

13. High-pitched bowel sounds (or none at all) are heard in patients with....

<u>Renal</u>

1. Lab Values

Specific gravity: 1.010 to 1.030

24-hour production: 1200 to 3000mL

BUN: 7- 18mg/dL

Creatinine: 0.7- 1.4mg/dL

2. Procedures

<u>Cystoscopy-</u>

Pre: Forced fluids, teach pt to deep-breathe to relax pelvic muscles

Post: check for bleeding

<u>IV Pyelogram-</u>

Pre: bowel preparation (eliminates bowel contents from blocking view of kidneys), NPO, check for allergies, burning normal during injection of dye, take xrays after dye
post: watch for dye reaction

<u>Ultrasound of bladder-</u> requires full bladder pre-procedure

<u>24-Hr Urine Collection</u> - collection of urine into one large container over 24 hrs

-discard first voided specimen at the start and time from there

-ice or refrigerate

-no preservatives in the urine container

3. Catheterization

<u>Female-</u> wipe from meatus toward rectum, insert 2-3 inches until urine returns, advance another inch and inflate balloon

<u>Male-</u> hold penis perpendicular to body and wipe meatus in circular motion

*replace foreskin after procedure

-sizes of catheters grow larger as the numbers growth larger

-typical catheter size is 16 or 18 French

When draining an extra-full bladder, only drain 500mL at a time.

Do not elevate bag above level of cavity.

4. Cystitis

<u>s/s</u>: urgency, frequency, burning during urination, cloudy foul smelling urine, confusion (elderly)

nrsg: force fluids to 3,000ml/day, antibiotics, encourage drinking cranberry juices, discourage caffeine, teach females to void post intercourse, wipe front to back, analgesics specific to urinary tract (see below)

5. Pyelonephritis

s/s: chills, fever, confusion (elderly) general malaise, flank pain, CVA tenderness

nrsg: IV fluids, antibiotics, bedrest during acute phase, encourage fluid intake 3000ml/day

Phenazopyridine	Genitourinary med, analgesic Antispasmodic	S/E: Bright orange urine
Oxybutynin	Genitourinary med, tx incontinence	S/E: Drowsiness, blurred vision

6. Renal and urethral stones

s/s: extreme flank PAIN, nausea, vomiting, hematuria, WBCs and bacteria in urine, low-grade fever and chills

nrsg: strain urine, recommend proper diet for prevention, analgesics, force fluids

7. Benign prostatic hyperplasia

s/s: retention, hesitancy, frequency, urgency, **nocturia, dribbling**, hematuria

labs: BUN, PSA

-PSA should be < 4ng/mL. If higher, needs to be investigated for BPH or possible CA

nrsg: TURP procedure, use CBI, I & O's, assess for hemorrhage, avoid long periods of sitting and strenuous activity until danger of bleeding is over

-Pink blood first day post-op TURP is okay, blood clots normal for first 24 hours

-if obstructed (s/s: bladder spasms, decr urinary output), turn off CBI and irrigate with 50mL, call physician if unable to dislodge

Finasteride	Tx of BPH (Proscar)	S/E: Decr libido, impotence, pregnant women should not come in contact with drug or pt's semen
Tamsulosin	Tx of BPH (Flomax)	S/E: Dizziness, headache, orthostatic hypotension Change position slowly, give at bedtime

8. Nephrotic Syndrome - kidney swollen

s/s: periorbital edema, pallor, lethargy, oliguria, dark frothy urine, decr serum protein, decreased albumin, elevated lipids, HTN

cause: prior kidney disease, diabetes, lupus

nrsg: steroid therapy and rest

9. Glomerulonephritis - infection/injury to glomeruli

s/s: hematuria, facial edema, fatigue, frothy urine, coughing, fluid in lungs, nocturia increased, nosebleeds, proteinuria, HTN

cause: usually related to strep infection (previous sore throat or impetigo (see peds))

nrsg: rest and antibiotics

10. Acute Renal Failure
a. Oliguric phase- <400mL/day, nausea and vomiting, drowsiness, hypertension

- protein sparing diet, restrict fluids, observe for hyponatremia and hypokalemia
b. Diuretic or recovery phase- 4-5L/day , incr serum BUN

- monitor I & Os

11. Ileal Conduit- surgical technique that diverts urine to exit the body through the ileum and into a bag

-mucus production is normal- if thick, encourage more fluids

-empty the bag frequently (max amt is ⅓ full), clean site with soap and water

-to prevent urine leakage when changing the bag, insert a gauze wick into the stoma

-use skin barrier to protect from acidic urine

-attach the appliance to a standard collection bag at night to prevent urine reflux into the stoma

12. Hemodialysis & Peritoneal Dialysis

Hemodialysis- AV fistula/AV graft

-check for thrill/ bruit every 8hr

don't use extremity for BP or blood specimens

Peritoneal Dialysis-

Peritoneal shunt, reposition if problems with outflow, clean catheter insertion site and apply sterile dressing, warm dialysate

Cloudy outflow- peritonitis

Outflow is normally clear to light yellow tinge

If a woman has red-tinged outflow, first ask if she has her period

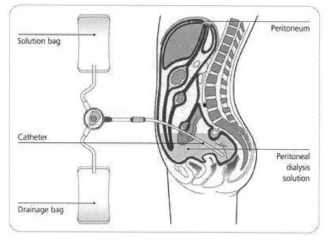

(picture credit: pathguy.com)

Renal Review Quiz

1. Normal BUN? Normal Creatinine? Normal specific gravity?

2. A patient reports burning during injection of dye during a pyelogram. What do you do?

3. When does CVA tenderness occur?

4. The TURP procedure is a treatment for...

5. Dark frothy urine is associated with _____ and is indicative of _____ in the urine.

6. Cloudy outflow during peritoneal analysis indicates...

7. With an AV fistula, you _____ the bruit and _____ the thrill.

8. What is oliguria?

9. Lab value to check for BPH is...?

10. An ultrasound for the bladder requires...

<u>Reproductive</u>

1. Menstrual cycle

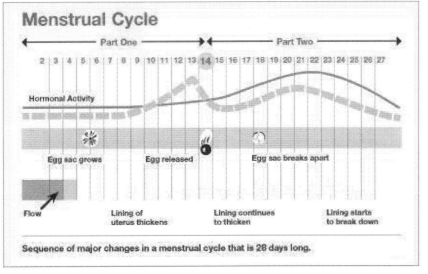

(picture credit: planned parenthood.org)

a. <u>Menstrual phase</u>- day 1-5- degeneration/discharge of tissue
b. <u>Proliferative phase</u>- day 6-14- follicle approaching maximum development in ovary
c. <u>Luteal phase</u>- Corpus luteum secretes progesterone, which changes the character of the uterine lining
d. <u>Ischemic phase</u>- corpus luteum degenerates; menstrual flow begins
   -during period, eat foods rich in iron, continue moderate exercise

Terms:

<u>Amenorrhea</u>- absence of menstrual flow

<u>Oligomenorrhea</u>- scanty flow

<u>Menorrhagia</u>- excessive flow, if pad is soaked in <2 hrs

<u>Dysmenorrhea</u>- painful menstruation

2. Ovulation

-typically occurs around day 14 in a 28 day normal cycle, pelvic discomfort is normal

-best method to detect is symptothermal- basal body temp increase + clear, slippery, stretchy cervical mucus

   -take body temp in the morning just before getting out of bed

3. Pap Smear

-advise no douching for at least 12h prior to test, sex is okay, have patient empty bladder just before exam, place patient in lithotomy position

4. Toxic Shock Syndrome- caused by using tampons

s/s: sudden onset of high fever, vomiting, diarrhea, drop in systolic blood pressure, diffuse sunburn-like macular red rash, later desquamation of palms and sole

-potential involvement with kidneys, cardiac, etc

tx: antibiotics, fluid and electrolyte replacement, education about tampon use

5. Uterine Fibroids- benign tumors of myometrium

s/s: low back pain, fertility problems, menorrhagia

nrsg: prepare for possible hysterectomy or myomectomy

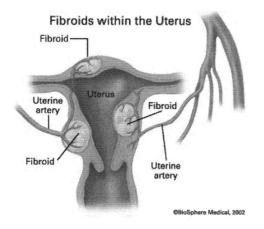

Fibroids within the Uterus

©BioSphere Medical, 2002

6. Uterine Cancer

s/s: watery discharge, irregular menstrual bleeding, menorrhagia

dx: endometrial biopsy

tx: radiation or hysterectomy

nrsg: **Internal Radiation Implants**

-enema, douche, low-residue diet, ample fluids

-indwelling catheter and fracture pan for elimination

-must restrict movements

-visitors/professionals limit exposure time, nurses cluster care

-no visitors <18 y/o, visitors <30 min/day and stay 6 ft away

-a lead shield is placed between the patient's bed and the hallway

-dislodged implant must be handled with special tongs by you, the nurse, and placed in lead-lined container for the radiation specialist to collect!

7. Ovarian Cyst

-pelvic discomfort, cramping pain

-palpable during routine exam

-will either burst by itself or need to be drained, may do biopsy

tx: Birth Control Pill (See below)

8. In Vitro Fertilization (IVF)

- in vitro fertilization, which means bypassing fallopian tubes

-removes ova via aspiration and mixes with sperm

-2 days later, fertilized ovum are returned to the uterus

-chance of multiple fetuses (twins, triplets, etc) is increased

9. Forms of Conception Prevention (Contraception)

Levonorgestrel/ Ethinyl Estradiol	Birth Control Pill *>1 pill missed requires another form of BC for rest of the cycle	S/E: Nausea, breakthrough bleeding, incr susceptibility to vaginal infections Adverse/Complication: DVT! Contraindications: HTN, thromboembolic disease, hx of circulatory disease, diabetes Take at the same time daily, no smoking. -antibiotics and antiepileptics interfere with BCP effectiveness
Levonorgestrel 75mg	Emergency birth control- "Plan B"	S/E: Nausea, breast tenderness, vertigo *most effect if taken IMMEDIATELY after unprotected sex, effect up to 12 hours later
Medroxyprogesterone	Depo Provera- given IM	S/E: irregular menstrual cycles and amenorrhea, weight gain, bleeding *admin q 3 months, increase calcium intake

Diaphragm- barrier device usually covered with spermicide

-do not douche

-inspect, wash with warm soap and water between uses

-do not remove sooner than 6 hours or later than 24 hours after sex

-change sizes if pt gains or loses > 20 lbs and also after childbirth

Abstinence/Calendar Method- abstaining around ovulation (days 10-14, 15-18)

-important to be accurate because sperm can remain viable 24-72 hours

Condoms- physical covering that fit over the penis

-leave space at the tip of the condom when fitted

-offer barrier against STI viruses

Natural skin type- do not protect against STI viruses

Surgery-

Vasectomy- minor procedure done under sedation, use contraception until declared sterile, easily reversed if desired

<u>Tubal ligation</u>- reversal not simple procedure, only 30% pregnancy rate after reversal

-may have sex 2-3 days after procedure

10. Breast Cancer

<u>s/s:</u> small immobile painless lump, change in color, nipple retraction

<u>tx:</u> surgery, radiation therapy or chemotherapy

Cisplatin, Platinol AQ, Methotrexate, Bleomycin	Antineoplastic (anti-all-cancers)	S/E: Bone marrow suppression, alopecia, renal toxicity, check CBC weekly, encourage fluids
Doxorubicin	Antineoplastic	S/E: red colored urine, extravasation of vein = necrosis

<u>nrsg:</u> advise **Self Breast Exam**:

-perform exam 1 week after the onset of each menstrual period (or routine time)

-inspect with arms at sides, arms above head, and hands on hips

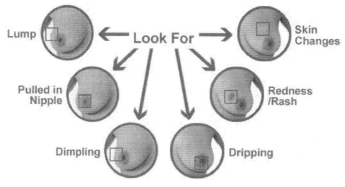

(picture credit: women-info.com)

11. Orchitis- swelling of the testes caused by mumps, bedrest, etc.

<u>s/s:</u> pain, swelling, could cause sterility

<u>nrsg:</u> ice packs to reduce swelling, scrotal support, bedrest

12. Testicular Cancer

<u>s/s:</u> painless testicular mass

<u>nrsg:</u> most common pt is males 20-34y/o, greatest risk is cryptorchidism

-should descend by 6 months

<u>cryptorchidism</u>- testicles undescended by 1 yr of age, fixed with orchiopexy

-perform testicular self exams regularly, in the shower where testes are relaxed

13. STD/STIs

-treat both partners

-female may not always show symptoms

-can easily become reinfected, though can easily be treated

-genital warts and genital herpes have higher risk for cervical cancer, also cannot be CURED, only treated with antivirals

Report to WHO: gonorrhea, syphilis, chlamydia

Do not have to report to WHO: HIV/AIDS, HPV, genital herpes, trichomoniasis

Reproductive Quiz

1. Define amenorrhea, oligomenorrhea, and menorrhagia.

2. Is sex okay before a pap smear?

3. What signs should be watched for as warnings of breast cancer?

4. Number one complication of the birth control pill?

5. Signs/symptoms of uterine cancer?

6. Guidelines for internal radiation implants?

7. What patient is most at risk for testicular cancer?

8. What is used to estimate ovulation?

9. A patient has an STD. Who else do you treat?

10. How can you cure warts and genital herpes?

<u>Maternity</u>

1. Placenta and Amniotic Fluid

<u>Placenta-</u> organ of pregnancy that permits an exchange, passes antibodies to fetus from mother, provides a barrier to some but not all harmful substances, provides nutrients

-perfusion influenced by maternal position, maternal BP, uterine contractions, condition of maternal blood vessels

<u>Amniotic fluid-</u> allows symmetric development, helps dilate the cervix once labor has begun, protects fetus from injury, keeps fetus at a stable temperature

-fluid pH = 7.1 to 7.3

-nitrazine strip turns blue if > 6, which would indicate ruptured membranes

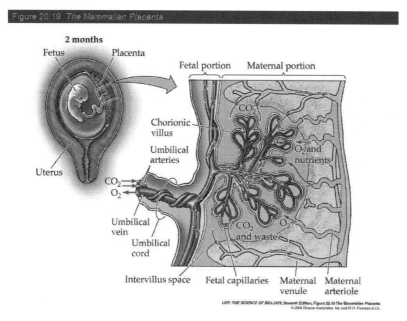

2. Fetal Development

4wks- heart formed and beating

12wk- all organ systems formed

16wk- external genitalia obvious, able to distinguish sex (quickening may occur)

20wk- fetal heartbeat can be heard, fetal movement should be felt by mother

25wk-28 wk- surfactant develops and lungs mature

32wk- most infants would survive if born, may need resp support

36wk- lanugo disappearing

40wk- vernix only in skin creases and folds

Preterm delivery is before 38wks

3. Pregnant Body Adaptations

Cervix-

Goodell's sign- softening

Chadwick's sign- blue-violet color

Hormones-

Lactogen- enhances milk production

Estrogen- hyperpigmentation, improves vascularization

Progesterone- relaxes muscles and ligaments throughout body (heartburn)

Body- total desirable weight gain is 23-28lb

-basal metabolic rate increases

-thyroid enlargement is normal

-increased vaginal discharge is normal; cheesy yellow-gray froth indicates infection

Fundal height- At umbilicus at 20 weeks

After 20 weeks, rises 1cm for every 1wk

Nutrition- extra 300kcal/day, should be no attempt at weight reduction during pregnancy

-should intake 600mcg/day of folic acid

Skin-

Striae gravidarum- pink or reddish streaks, stretch marks

Chloasma- "mask of pregnancy", incr pigmentation can occur

Linea nigra- dark line from symphysis pubis to umbilicus

Breast soreness- support with well-fitting supportive bra

Striae gravidarum	Chloasma	Linea nigra

Edema- ankle edema related to decr venous return from lower extremities

Nrsg: put feet up, med management if shows warning signs of preeclampsia

Vital Signs-

-Incr baseline HR, NOT blood pressure- warns of pre-eclampsia

Digestion-

-morning sickness- nau and vom due to hormonal levels, recommend to eat crackers ½ hr to getting up in the morning or eating a meal, small amounts of bland food/milk, consume most liquids between meals

Pica- food cravings for unusual substances

Exercises-

-pelvic rocking helps to relieve backache and pain during early labor by stretching muscles
-deep breathing for relaxation
-sitting and squatting help stretch the perineal muscles

4. Verifying Pregnancy

Presumptive- suspicion, not proof: amenorrhea, nausea/vom, breast sensitivity, fatigue/lassitude, quickening

Probable- incr suspicion but still no proof: uterine enlargement, positive urine pregnancy tests (rely on HCG)

Positive- definite signs of pregnancy: fetal heartbeat, palpation of fetal movement, outline of fetal skeleton by sonogram or xray

5. Naegele's rule

-add 7 days to the first day of the last menstrual period, subtract 3 months and add 1 year

6. GTPAL

Gravida- total number of pregnancies regardless of duration (including this one)

Term- fetuses delivered at term (completion of 37 weeks/beginning of 38)

Preterm- fetuses delivered at 20 weeks to completion of 37 weeks

Abortion- spontaneous or elective abortions before 20 weeks

Living Children- living children

7. Tests

Amniocentesis- indicated 14-17wks to determine chromosomal abnorms

Indicated after 28 wks to determine sex and sex-linked disorders

Indicated after 30 wks to determine lung maturity

-pt should empty bladder prior

Chorionic villus sampling- diagnoses genetic disorders from 8-12 weeks gestation

-pt should have full bladder prior

Nonstress Test (NST)- evaluates FHR in response to monitored fetal movement, woman should eat 2hr before

       -Reactive Result- FHR accelerations, good!

       -Nonreactive Result- not enough accelerations or presence of decelerations, bad!

Contraction/Stress Test (CCT/OCT)- evaluates FHR in response to contractions initiated by IV Pitocin (used to use nipple stimulation!), not performed prior to 28 weeks

       -Negative Result- No late decelerations, good!

       -Positive Result- Late decelerations with at least 50% of contractions, bad!

Biophysical Profile- an ultrasound that evaluates fetal breathing, movements, fetal body tone, amniotic fluid volume, fetal heart rate reactivity

-movements: good if moves more often than 3 times in 1 hour, min of 10 times per day

Group B Strep (GBS)- swab of vagina and rectum at 36 weeks

tx: IV PCN during labor

8. Braxton-Hicks contractions- weak, unorganized contractions, stop with rest, change of position or change of activity

True labor- contractions that are regular with shortening intervals, incr in duration and intensity, contractions do not decr with rest, cervix progressively effaced and dilated
-call health care provider when contractions occur every 5 minutes for an hour

9. Bleeding in Pregnancy

First Trimester-

-"spotting" with no pain = completely normal

nrsg: limit activity until stopped

-"spotting" with cramping pain = threatened abortion, requires f/u

Ectopic pregnancy- implantation outside uterus, potentially life-threatening

s/s: Unilateral lower quadrant abdominal pain, low HCG levels, rigid and tender abdomen, signs/symptoms of hemorrhage, Cullen's sign

nrsg: monitor Hb and Hct, prepare for surgery, adequate blood replacement

Placenta previa- placenta is low-lying or blocking cervix

s/s: painless frank red vaginal bleeding

nrsg: bed rest or modified Trendelenburg, home if bleeding ceases and activity limits, no coitus, possible delivery by c-section

Abruptio placentae- placenta prematurely separates from uterus

s/s: abdominal pain, dark red vaginal bleeding

nrsg: immediate delivery or c-section, manage hemorrhage

Hydatidiform mole- degenerative anomaly- results in molar tissue that can become malignant

s/s: Nausea, vomiting, hypertension (if bleeding), hypertension possible (r/t edema), greatly elevated HCG levels, no FHR, uterine size greater than expected, minimal vaginal bleeding with passage of grapelike clusters, brown vaginal discharge with "clear vesicles"

tx: curettage to completely remove all molar tissue

*pregnancy discouraged for 1yr and HCG levels are monitored. If they remain elevated, may need hysterectomy

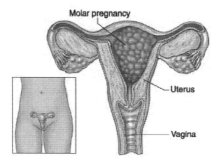

(image credit: mdguidelines.com)

10. Maternal Hypotension

-could be due to painkillers, epidural anesthesia or supine positioning (avoid!)

Mild/moderate- place mother in left lateral position, incr rate of IV fluid, admin O2 by mask, stop Pitocin

Severe- place mother in Trendelenburg position for 2-3 min

11. Diabetes in Pregnancy

-if pt is already diabetic, great diabetic control will be needed. Insulin needs will incr with pregnancy and decr to original levels after delivery

-gestational diabetes- normally controlled only via diet, do not use oral hypoglycemic

nrsg: walking is excellent exercise

-diabetes screening occurs at 24-28 wks for all gravidas

12. Rhogam

-only administer Rhogam to an Rh- woman who has delivered her first Rh+ baby

-given at 28 weeks and within 72 hours after delivery, IM in dorsogluteal

-indirect Coombs test measures antibodies in the blood of mom

-no point in administering to an Rh- woman who is delivering her second Rh+ baby because the antibodies have already been formed

Rho(D) immune globulin	Immunoglobulin	S/E: Back pain, irregular heartbeat, diarrhea, nausea, vomiting, rapid weight gain

13. Pregnancy-Induced Hypertension

Pre-eclampsia- elevated blood pressure, edema (especially facial), proteinuria

HELLP syndrome- **H**emolysis, **e**levated **l**iver **e**nzymes, **l**ow **p**latelet count

Eclampsia- when tonic-clonic generalized seizures, coma, HTN crisis or shock occurs

nrsg: Bedrest side-lying, monitor kidneys, seizure precautions, antihypertensives and anticonvulsants, monitor FHR and maternal status

Magnesium sulfate/chloride	Anticonvulsant, used for pre-eclampsia. May also be used as a supplement. level = 4 - 7, > 8 is toxic. antidote: calcium gluconate	S/E: Flushing, thirst, hypotension, check knee jerks before each dose. A/E: Resp depression

14. Fetal Heart Rate

Normal- 120-160BPM

Tachycardia- >160BPM

Bradycardia- <110BPM

V	C
E	H
A	O
L	P

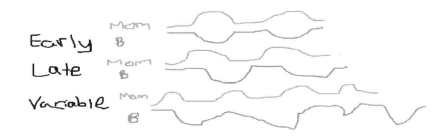

Early   Mom B

Late   Mom B

Variable Mom B

Accelerations- rise above baseline in response to fetal movement or contractions

Early Decelerations- occurs before peak of contraction, "mirror image", associated with head compression and is okay

Late Decelerations- occurs after peak of contraction, slow to return to baseline, indicates fetal hypoxia, stop pitocin

Variable Decelerations- transient reduction occurring at any time, indicative of cord compression, which may be relieved by a change in the mother's position, stop pitocin

15. Fetal Lie & Presentation

Cephalic- head presentation- usually occiput (O)

Breech- rear or feet presentation- sacrum (S)

Shoulder- transverse lie- scapula (SC)

Expressed as LOA = left occiput anterior

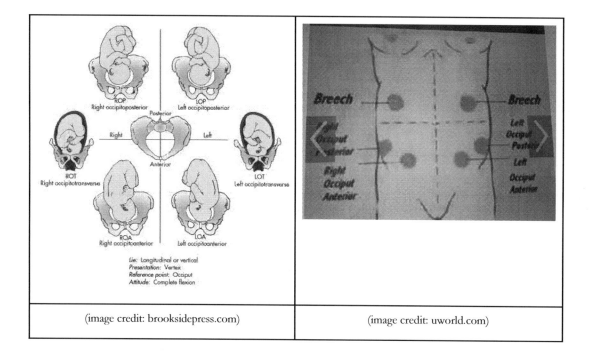

| (image credit: brooksidepress.com) | (image credit: uworld.com) |

Station: -5 to -1 indicates a presenting part about zero station (floating), +1 to +5 indicates a presenting part below zero station

Zero station = ischial spines of maternal pelvis

Leopold's maneuvers- four systematic movements by health practitioner meant to determine the position of a fetus inside the mother

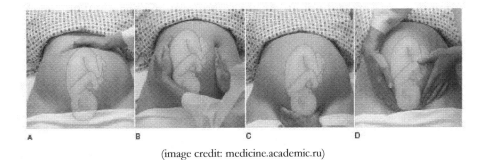

(image credit: medicine.academic.ru)

16. Stages of labor

Onset- Check FHR by auscultation for 1 min and with next contractions

-expulsion of mucus plug

-may be a gush or trickle of fluids- questionable should be tested for alkalinity

-premature rupture of membranes is associated with chorioamnionitis (infection)

-for a prolapsed cord, place client in knee-to-chest or Trendelenburg position, do not attempt to replace cord, if you can see fetal head, put gentle upward pressure on it

First Stage: From onset of regular contractions to full dilation

    a.   Latent phase- 0-3cm
    b.   Active phase- 4-7cm
    c.   Transition- 8-10cm, contractions every 2-3min, last 60-90sec

DO NOT ENCOURAGE PUSHING UNTIL FULLY DILATED

-normal progress for primipara: 1-1.2 cm per hour, up to 3 hours of pushing

-nrsg: recommend ambulation, slow chest breathing, effleurage

Pitocin	Labor drug to increase contractions	S/E: Late decelerations, variable decelerations, hypertonic uterine patterns ANY of above? Stop Pitocin

Second Stage: Compete dilation to delivery of the baby

-bearing down/pushing incr intra-abdominal pressure, client may feel as though she needs to use the bathroom

presenting part descends, crowning occurs

-cardinal movements during delivery: descent, flexion, internal rotation, extension, external rotation, expulsion

-mother may need episiotomy- surg incision of perineum to prevent tearing

Third Stage: Birth and delivery of placenta

    a. Separation of placenta from the uterine wall- most reliable sign is sudden lengthening of cord
    b. Expulsion of placenta through the vagina
    c. Contraction of the uterus controlling hemorrhage and producing placental separation

Fourth Stage: Immediate recovery period from delivery of placenta to stabilization of maternal systemic responses and contraction of the uterus

-total blood loss should be around 500mL, >1000mL indicates hemorrhage

-heavy if > 1 pad per hr

17. APGAR scores, at 1 and 5 minutes to evaluate condition at birth

-stands for appearance, pulse, grimace, activity and respirations

	0	1	2
Appearance	Blue, pale	Body pink, extremities blue	Completely pink
Pulse	Absent	Slow <100	>100
Grimace	No resp	Weak cry	Vigorous cry
Activity	Flaccid	Some flexion	Active
Resp effort	Absent	Slow, irregular	Good cry

18. Postpartum Body Norms

Vital Signs:

-bradycardia and slightly increased temperature (up to 100.4 degrees F) are normal up to 24 hr after delivery

Fundus: at umbilicus for first 12hr, then descends by 1cm each following day

-if it feels boggy, nursing intvn is to massage it

-if it is deviated from midline, encourage pt to empty bladder or catheterize

Lochia:

Rubra- day 1-3

Serosa- day 4-9

Alba- day 10+

Breasts, Nipples, & Lactation:

Non-nursing woman: suppress lactation with a tight-fitting bra, cold cabbage leaves, ice packs, minimize breast stimulation

Nursing woman: depends on infant sucking and let-down reflex

Nipple care: clean with water, no soap, dry thoroughly, expose to air

Plugged duct: relieved by heat and massage

Mastitis: continue breast-feeding, take antibiotics, use warm compresses

During nursing: fully empty one breast per feeding, then empty other on next feeding

-need an extra 500 calories daily

-less insulin when breastfeeding

-nurse 15-20 minutes at each breast

-feed q2-3 hours, q4 hrs at night

19. Postpartum blues are normal for the first week, report if prolonged

20. Sexual activities okay once lochia has ceased (3-4 weeks)

Breastfeeding is not a reliable method of birth control

No hormonal contraceptives (birth control pill) should be used during breastfeeding

Kegel exercises- strengthens pelvic and pubococcygeal muscles

21. Disseminated Intravascular Clotting (DIC)

-a state of diffuse clotting in which clotting factors are consumed, leading to widespread bleeding

-can occur with intrauterine fetal death, abruptio placentae, septic shock, etc

s/s: petechiae, oozing from injection sites, and hematuria

tx: give supportive care, fluids

22. TORCH Infections and Pregnancy

-TORCH = toxoplasmosis, other (syphilis, varicella, etc), rubella, cytomegalovirus, herpes

Toxoplasmosis: found in cat feces, AVOID cats

Rubella: pregnant woman cannot get MMR vaccine

Herpes: can be transmitted to the infant during a vaginal delivery

      -c-section is warranted if the mother has active lesions and should occur before ruptured membranes or any active labor

23. HIV and Pregnancy

-mother should receive HIV medicines during pregnancy and birth process

-may have a vaginal or c-section depending on viral load

-oral zidovudine given to child after birth

-mother may not breastfeed

-transmission of HIV virus from mother to baby with these precautions is VERY low, but all infants will test positive for HIV at birth due to the mother's antibodies

Zidovudine	Treats HIV	S/E: Headache, decreased CBC

Maternity Review Quiz

1. When is the heart formed?

2. When does quickening occur?

3. Define a preterm delivery.

4. What is Goodell's sign? Chadwick's?

5. What is the approximate fundal height of a mom 30 weeks pregnant?

6. How many extra calories are recommended per day for a pregnant woman?

7. What is normal ankle edema? What is abnormal?

8. What VS does increase during pregnancy? Which does not?

9. What is Naegle's rule?

10. An amniocentesis at 32 weeks is used for what?

11. Prep for a chorionic villus sampling is what?

12. A reactive result for a NST is…

13. Unilateral lower quadrant pain and heavy bleeding could be a sign of…

14. Painless frank bleeding is a sign of...
    Painful bleeding/diffuse abdominal pain is a sign of...

15. Elevated HCG levels + no FHR + fundal height greater than expected =....

16. Insulin needs _____ with pregnancy.

17. Three signs of pre-eclampsia?

18. Antidote to magnesium sulfate?

19. Where is zero station?

20. Describe what care a mother with HIV should take.

21. How can a pregnant woman avoid toxoplasmosis?

The Neonate

1. VS of Neonate

-normal weight loss of 5-10% during first few days, should be regained in first week

-breastfed babies should surpass birthweight at 10-14 days with 6-8 wet diapers per day

-normal for both male and female babies to have swollen genitals and breast tissues d/t maternal hormones

temperature: 97.7-98.6 axillary

apical HR- 100BPM in sleep, 120-140BPM awake, up to 180BPM crying

respirations- 30-60/min, up to 15 sec of apnea is normal, irregular

blood pressure- 65/41mm Hg

blood sugar - 40 - 60 via heel prick

2. Skin Variations

acrocyanosis (bluish discoloration of hands/feet) is norm for 1st 24hr after birth

molding- superficial scalp changes due to vaginal delivery, disappears within a few days

vernix caseosa- protective fatty substance, do not attempt vigorous removal

lanugo- fine down of hair, extensive amount is indicative of prematurity

Mongolian spots- bluish gray or dark pigmentation on lower back/buttocks

telangiectatic nevi- small, flat red localized areas of capillary dilation- fade during infancy

nevus vasculosus- raised, demarcated, dark red, rough-surfaced hemangioma, grows rapidly for several months and then fades, disappears by 7y/o

nevus flammeus (port wine stain)- reddish, usually flat discoloration commonly on the face/neck, does not grow or fade

epstein's pearls - white bumps in infant's mouth - cysts that are totally normal when present at birth

milia - baby acne

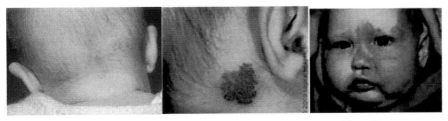

| Telangiatic nevi | Nevus vasculosus | Nevus flammeus |
| (picture credit: webmd.com) | (picture credit: quizlet.com) | (picture credit: quizlet.com) |

anterior fontanelle- diamond-shaped, closes by 18mos

posterior fontanelle- triangular, closes between 8-12wks of life

3. Neonate Preventative Treatments

Vitamin K shot – given because newborns do not have the bacteria to make their own yet

Antibiotic eye drops – given to prevent eye infections r/t gonorrhea/bacteria

PKU test - blood sample taken at birth, then repeated, to test for a missing liver enzyme (recessive genetic disorder) that would cause permanent brain damage with undigested proteins

-PKU test known as Guthrie test - most reliable after ingestion of protein

- if positive, neonate needs special formula called Lofenlac and stays on protein-free diet for life

4. Umbilical Cord & Circumcision Care

cord –shriveled and black by 2-3 days, falls off within 1-2wk, foul-smelling d/c indicates infection

circumcision- post-procedure, leave yellowish exudates ON, do NOT clean off, teach to clean by squeezing warm water over penis and dry gently, vaseline between penis and diaper

5. Reflexes

-palmar grasp- fades by 3-4 mos

-tonic neck- fades by 3-4 mos

-moro- startle reflex, fades by 3-4 mos

-stepping reflex- fades 4-5 mos

-rooting/sucking- fades by 4-7 mos

-Babinski's sign- fades by 12 mos

6. Cold Stress/ Hypoglycemia

s/s: mottling of skin/cyanosis, jitteriness, weak high-pitched cry, lethargy, eye-rolling

nrsg: admin glucose, warm

7. Hyperbilirubinemia

a. Physiologic jaundice- normal after first 24h of life, disappears within 5-7 days

b. Pathologic jaundice- occurs within first 24h of life, lasts longer than 7 days in full-term or 10 days in preterm infants

tx: phototherapy or exchange transfusions

-expect jaundice around eyes to disappear last, cover eyes and genitals

-expect jaundice to last longer in breastfed babies

-expect increased urinary output and watery, green stools

-check vital signs every 2-4 hours

-lab technician should turn off lights before drawing labs

8. Neonatal Abstinence Syndrome

<u>s/s</u>: high-pitched cry, irritable, hyperreflexivity, decr sleep, diaphoresis, tachycardia, vomiting, uncoordinated or weak sucking

<u>nrsg</u>: reduce environmental stimulation, provide adequate fluids/nutrition, assess muscle tone

9. Retinopathy of prematurity

-blindness in premature infants caused by high concentrations of oxygen

Neonate Review Quiz

1. What is a normal amount of weight loss for a neonate?

2. What is vernix caseosa?

3. When does the posterior fontanelle close?

4. When does the stepping reflex fade?

5. Name signs of cold stress.

6. What is pathologic jaundice?

7. What is a high-pitched cry a sign of?

8. Intervention for jaundice in neonates?

9. What is retinopathy of prematurity caused by?

10. When does Babinski's sign fade?

Pediatrics

1. Theories of Development

<u>Erickson's Psychosocial Stages</u>

Age	Crisis
1st yr of life	Trust vs. Mistrust
2nd yr of life	Autonomy vs. Doubt
3rd through 5th yrs	Initiative vs. Guilt
6th yr to puberty	Industry vs. Inferiority
Adolescence	Identity vs. Role Confusion
Early Adulthood	Intimacy vs. Isolation
Middle Age	Generativity vs. Self-Absorption
Aging Years	Integrity vs. Despair

2. Developmental Standards

<u>Standardized Growth Chart</u>- best estimation of a child's height/growth as average or abnormal

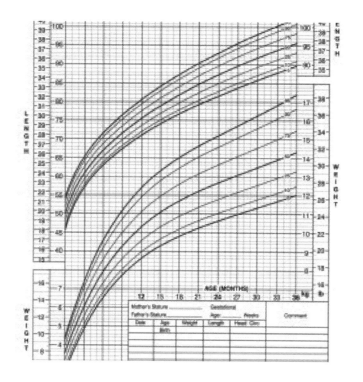

<u>Denver II</u>- evaluates children from birth to 6y in personal-social, fine motor, language, gross motor

3. Age-Appropriate Prep for Medical Procedures

6-12 mos- model desired behavior

Toddler- simple explanations, use distractions, allow choices but use consistent rituals

Preschool- Encourage understanding with puppets/dolls, demonstrate equipment

School Age- Allow questions, explain why, allow to handle equipment

4. Milestones

1 mos- head sags, early crawling movements

2 mos- closing of posterior fontanelle, able to turn from side to back, eyes begin to follow a moving object, social smile appears

3 mos- can bring objects to mouth at will, head held erect

4 mos- pleasure in social contact, drooling

5 mos- birth weight doubled

6 mos- teething begins, early ability to distinguish strangers

7 mos- sits for short periods using hands as support, lability of mood

8 mos- anxiety with strangers, stands up

9 mos- elevates self to sitting position, one word expressions (dada)

10 mos- crawls well, pulls self to standing position with support

12 mos- birth weight usually tripled, sits from standing position without assistance, 3-5 words

15 mos- walks alone, builds 2-block tower, throws objects

18 mos- anterior fontanelle closed, climbs stairs, oral vocab 10 words

24 mos- early efforts at jumping, 5-6 block tower, 300-word vocab, obeys easy commands

30 mos- 7-8 block tower, stands on one foot, has sphincter control for toilet training

3 yrs- learns from experience, rides tricycle, walks backward, undresses without help

4 yrs- climbs and jumps well, laces shoes, brushes teeth, throws overhead

5 yrs- runs well, jumps rope, dresses without help, begins cooperative play, ties shoes

6 yrs- show-off, self-centered, sensitive to criticism of others, ties knots

7 yrs- team games/sports/organizations, concept of time

5. Discipline

– approximately 1 minute in time out for each year of the child's age (4 min for 4 y/o)

-Authoritative parenting style works best- limits set consistently with atmosphere of open discussion for the child

6. Poison Control/Prevention

-initiate steps to stop exposure

-if conscious- call poison control, follow their directions

-more toxic to inhale a poison than to absorb it through skin or digest it

-activated charcoal only works if given within 1hr of ingestion

-for aspirin OD, greatest risk is bleeding-> induce vomiting

Acetaminophen	Pain/fever medication	Adverse eff: Liver damage
		Antidote: N-acetylcysteine, must be administered within 8 hours, wait 4 hours for levels
		Max adult dose = 4g/24 hrs

Lead toxicity- anemia, hearing impairment, distractibility, incr intracranial pressure, seizures- admin chelating agent, give milk

-children get a lead level drawn at 1 yr of age

7. Immunizations -mild fever/aches treated with Tylenol are expected, but not very high fever or severe illness

# of Shots	Immunizations	2 months	4 months	6 months	1 yr	4-6 yrs	11-12 yrs
1	Varicella				#1		
1	Tetanus booster						#1
1	Meningococcal						#1
2	MMR				#1	#2	
3	Hepatitis B	#1	#2	#3			
4	Hib	#1	#2	#3	#4		
4	IPV	#1	#2	#3		#4	
4	PCV	#1	#2	#3	#4		
5	DTAP	#1	#2	#3	#4	#5	

-live vaccines = varicella and MMR

-minimum age for flu shot is 6 months

8. Pyloric Stenosis- obstruction of passageway from stomach to intestines

s/s: projectile vomiting with streaks of blood in it, weight loss, constipation, dehydration, olive-sized tumor

nrsg: pre-op, position upright after small feedings, keep quiet environ after feeding, take weights to watch for dehydration, K+ is expected to be LOW

post-op, small freq feedings of glucose water or electrolyte solution 4-6hr post op

risk factor: maternal polyhydramnios

9. Cleft lip and palate

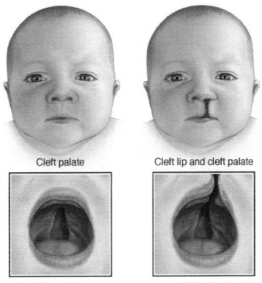

<u>Cleft lip</u>- repaired during first weeks of life

<u>Palate</u>- repaired between 12-18 mos

nrsg; provide support to parents, assess infant's ability to suck, do not allow infant to suck after repair until it has healed

10. Esophageal atresia and tracheoesophageal fistula

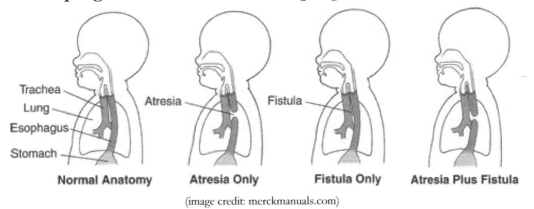

(image credit: merckmanuals.com)

<u>s/s</u>: choking, coughing, cyanosis, excessive saliva, excessive mucus, stomach distention, substernal retractions

<u>nrsg</u>: prevent saliva aspiration, post-op: care for incision line and provide TPN until oral feedings tolerated, support parents

risk factor: maternal polyhydramnios

11. Hypospadias / Epispadias

-condition in which the urinary meatus is located on the ventral/dorsal surface of the penis

-surgically corrected between 6-12 months of age

-child is not circumcised until after the surgery in order to use the extra skin for reconstruction

-catheterized post-op 1-2 weeks

12. Congenital Heart Anomalies

s/s: cyanosis, clubbing of fingers, marked exercise intolerance, difficulty eating, small stature, failure to thrive

nrsg: optimal nutrition, small frequent feedings, much rest, provide a calm environment, enlarged nipple hole, knee-to-chest position, have child blow around thumb with closed mouth to valsalva

13. Spina bifida/neural tube defects

s/s: bulging, sac-like lesion exposed to air

nrsg: position on abdomen, cover lesion with moist sterile dressing, identify possible motor and sensory dysfunctions

tx: post-op, parents must know that any damaged function cannot be fixed.

14. Cerebral palsy- a neuromuscular disability in which the voluntary muscles are poorly controlled because of brain damage

s/s: neonate cannot hold head upright, feeble cry, weakness, failure to thrive, signs of mental retardation

nrsg: assist with feeding, never tilt head backward when feeding, admin high calorie diet, given Baclofen

Baclofen	Tx muscle spasticity	S/E: Muscle weakness, seizures, nausea, constipation

15. Developmental dysplasia of the hip

s/s: uneven gluteal folds and thigh creases, limited abduction of hip, prominent trochanter, short limb on affected side

dx: x-ray evaluations, Ortolani maneuver

nrsg: splinting, or hip spica cast in severe cases

Pavlik harness- kept on 23 hours per day, do frequent skin checks

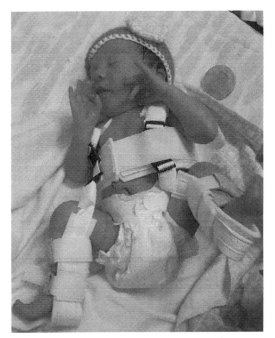

(image credit: Lisa Chou- above is her daughter Elena @ 1 week old)

16. ADHD/ADD

s/s: shortened attention span, poor grades in school, inability to sit still or focus on activities

nrsg: collaborate with teachers for IEP in school, may need to give 2pm dose

-meds have a paradoxical effect in this condition despite being stimulants

Methylphenidate	CNS stimulant Monitor CBC, give at least 6 hours prior to bedtime	S/E: Lightheadedness, nervousness, tachycardia, weight loss
Dextroamphetamine	CNS stimulant Give 6 hours prior to bedtime	S/E: Tachycardia, weight loss, agitation

17. Scoliosis

s/s: poor posture, uneven hips or scapulae, kyphosis, uneven waistline

-Milwaukee brace- effective for 30-40 degree curves (mild to mod), 4-6 yr program

-wear for 23.5 hours daily with ½ hour for personal hygiene, wear protective shirt under the brace, skin care to pressure areas, isometric exercises to strengthen abs

18. Clubfoot- involves abnormality of talus bone

-frequent casts are attempted before surgery

19. Increased ICP

s/s: Bulging fontanelles, high-pitched cry, lethargy, seizures, poor appetite

20. Hirschsprung's Disease- lack of nerves at the anus so lack of peristalsis

s/s: failure to pass meconium shortly after birth, infrequent but explosive stools, jaundice, poor feeding, poor weight gain, vomiting, diarrhea

dx: rectal biopsy

tx: surgery to remove aganglionic portion of bowel and temporary colostomy

21. Croup

Def: baby version of a cold, only with tiny airways and no way to blow their nose

s/s: hoarse or "barking" cough, fever

Precaution: CONTACT!

Tx: stand in bathroom with hot shower running and steam or stand in front of open freezer, both aid breathing

-if pt begins drooling and inspiratory stridor- epiglottitis. Do not look down throat! Keep an emergency tracheostomy kit in the room

22. Impetigo- highly contagious bacterial skin infection (staph) most common among preschool children

s/s: honey-colored scabs formed from dried serum on face, red blisters on face

tx: bacterial ointment, wash with soap and water

cx: can develop into glomerulonephritis

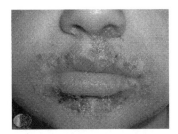

(picture credit: doctorgrasshopper.wordpress.com)

23. Critters

Lice- known as pediculosis capitis
s/s: features itching
-spread by close contact
-can only live 48 hours without host
-wash clothing, linens, combs, etc in hot water and dry with high heat
-nonwashable items should be sealed in a plastic bag

| Lindane | Tx for lice "Kwell" | S/E: nausea/vomiting-> overapplication |
| | | Apply and inspect hair shafts, check for nits daily x 1 week |

Pinworm- emerges from the rectum at night to lay eggs onto the perineal area
s/s: itching around buttocks
-spread fecal-oral route
Tx- Handwashing, keep fingernails short, see mebendazole

| Mebendazole | Treats worm infections | S/E: Rash, nausea, vomiting, diarrhea |

Ringworm- fungal infection known as tinea corporis
s/s: circular, scaly patches with or without areas of alopecia
-spread through contact

| Clotrimazole, miconazole | Antifungal medication, topical | S/E: redness, allergy |
| | | -use for 1-2 weeks after resolution |

24. Tonsillectomy

-watch out for signs of excessive bleeding- drooling, bright red blood, frequent swallowing
-liquids and soft food diet, but NO red foods, NO milk and milk products, NO citrus or carbonated liquids
-best position is side-lying
-watch out for presence of loose teeth prior to surgery

Pediatrics Review Quiz

1. When does the identity vs confusion stage take place in Erikson's theory?

2. When is the latent stage?

3. When can a baby sit without support? When can the child build a two block tower?

4. What does the Denver II evaluate?

5. What ages are appropriate to use puppets/dolls to explain procedures?

6. When are the Dtap immunizations given to a child?

7. Antidote to a Tylenol overdose?

8. Biggest symptom of pyloric stenosis?

9. Cleft lip repaired when? Cleft palate?

10. Important psychosocial teaching for parents of kids with spina bifida?

11. Uneven gluteal folds are a sign of...

12. Treatment for Hirschsprung's disease?

13. Two aids for breathing for a child with croup?

14. What can impetigo turn into?

Genetics

1. Down's Syndrome
- caused by trisomy 21, associated with maternal age > 35

s/s: mental deficiencies, marked hypotonia, short stature, epicanthal folds, low set ears, protruding tongue, low nasal bridge

nrsg: provide stimulation (PT, OT), observe for signs of common physical problems- heart disease, hearing loss, celiac disease

2. Huntington's Disease
- caused by inherited degenerative disease where a protein called "huntingtin" mutates and damages cells

type: autosomal dominant

s/s: depression and temper outbursts, choreiform movements, personality changes, paranoia, memory loss, decr intellectual function, dementia

nrsg: tx is symptomatic, drug therapy to reduce extra movement and subdue behavior changes

3. Amyotrophic Lateral Sclerosis
- caused by inherited degenerative disease involving the lower motor neurons of the spinal cord

type: autosomal recessive, autosomal dominant, or x-linked are all possible

s/s: tongue fatigue/atrophy with fasciculations, dysphagia, muscular wasting that begins in upper extremities

tx: psychosocial management, treat self-care deficits

4. Parkinson's Disease
- caused by inherited degenerative disease of dopamine-producing neurons in the substantia nigra of the midbrain

type: autosomal recessive

s/s: tremors (pill-rolling), bradykinesia, restlessness, rigidity and propulsive gait, weakness, masklike facial expression, cognitive disturbances

nrsg: promote comfort and rest, keep pt as functional as possible, teach ambulation modification, thickened liquids are better to decrease the risk of aspiration, swing arms when walking, exercise early in the day

Benzatropine	Used to reduce symptoms of Parkinson's (Cogentin)	S/E: orthostatic hypotension, taper before d/c
Levodopa	Antiparkinsonian	S/E: Mental status changes, twitching, anorexia Administer in long, prolonged doses
Carbidopa/Levodopa	Antiparkinsonian	S/E: Hemolytic anemia, dystonic movements, dry mouth

5. Duchenne Muscular Dystrophy
- progressive hereditary muscle disorders characterized by muscular weakness

type: x-linked recessive

s/s: leg weakness, stumbling and falling, pseudohypertrophy of muscles, lordosis, waddling gait, contractures of elbows, knees, feet

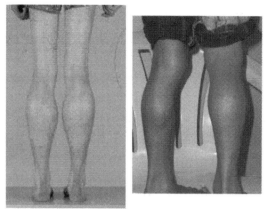

(image credit: slideshare.net)

<u>dx:</u> CPK (muscle enzyme) elevated, abnorm muscle biopsy
<u>nrsg:</u> promote safety, emotional support, prevent contractures, discuss balance between activity and rest

6. Cystic Fibrosis- hereditary defect in the exocrine system characterized by thick secretions and problems with digestion
<u>type:</u> autosomal recessive
<u>s/s:</u> thick mucus, inability to digest foods well, poor weight gain, chronic lung infection, SOB, difficulty in becoming pregnant, more likely to experience pancreatitis and diabetes
<u>dx:</u> sweat chloride test
<u>nrsg:</u> use pancrelipase with food to help digest, postural drainage and RT
<u>cx:</u> steatorrhea, infants have incr risk for intussusception

7. Sickle Cell Disease
<u>problems:</u> hydration, oxygen, pain

<u>type:</u> autosomal recessive, most common in pts of African descent
<u>nsg:</u> oxygen therapy, bedrest, fluids, pain management, Hydroxyurea (antisickling agent), avoidance of stressors that could trigger a worsening (high altitude, dehydration, strenuous exercise, tight clothes)

8. Hemophilia A
<u>s/s:</u> spontaneous easy bruising, joint pain with bleeding, pallor

<u>type:</u> x-linked recessive
<u>tx:</u> transfusions, analgesics
<u>nsg:</u> analgesics, avoid IM injections, no aspirin, stop topical bleeding with ice, pressure, replacement of deficient clotting factors, physical therapy after bleeding episodes, no contact sports
<u>cx:</u> joint stiffness/arthritis, hemorrhage

Factor VIII	Missing factor in hemophilia A	Complication: can develop antibodies forming against it and subsequent reactions

106

Genetics Review Quiz

1. Down's syndrome is associated with what risk factor?

2. Huntington's disease includes dementia? T/F?

3. ALS progression occurs over how long? Starts where?

4. Is intellect impaired in Parkinson's?

5. Signs and symptoms of muscular dystrophy?

6. What is a complication of hemophilia?

7. What would dehydration do to a sickle cell patient?

8. What degenerates in ALS?

Session 8: Autoimmune Disorders

1. Lupus erythematosus- autoimmune condition in which the body attacks its own healthy tissues and causes damage

s/s: arthritis, butterfly rash, fatigue, fever, hair loss, mouth sores, swollen lymph nodes

tx: immunosuppressive drugs: corticosteroids and hydroxychloroquine, avoid sun

-watch for proteinuria and hyperlipidemia in labs

Dexamethasone	Glucocorticoid, corticosteroid, immunosuppressant	S/E: Edema, osteoporosis, hyperglycemia Do not abruptly d/c, check weight, protect from fractures.

2. Myasthenia gravis- caused by an autoimmune process where there is a deficiency of acetylcholine at myoneural junction because it has been destroyed

s/s: extreme skeletal muscle weakness, quickly produced by repeated movement but disappears following rest, diplopia, ptosis, impaired speech, choking/aspiration of food, respiratory distress
dx: "Tensilon" test, nerve conduction study, electromyography (EMG)
nrsg: promote comfort, rest, relieve symptoms by administering drugs

Pyridostigmine	Parasympathomimetic, used to tx myasthenia gravis	S/E: Sweating, diarrhea, nausea, increased bronchial secretions, facial flushing *Doses MUST be given on time *If undermedicated, can prompt a myasthenic crisis-> respiratory paralysis, muscle weakness *If overmedicated, can prompt a cholinergic crisis-> respiratory paralysis, muscle weakness
Edrophonium	"Tensilon"	-used to diagnose myasthenia gravis

3. Multiple Sclerosis- chronic progressive autoimmune disease characterized by demyelination and scarring throughout the brain and spinal cord

s/s: ataxia, weakness, nystagmus, chewing and swallow difficulties, paresthesias, incontinence/retention of urine & constipation, emotional lability, sexual impairment
-features Charcot's Triad - nystagmus, intention tremor, and dysarthria

nrsg: encourage fluid intake 2000mL/day, progressive resistance exercises, ROM, bladder and bowel training, nutritional therapy

4. Guillain-Barre syndrome

def: an autoimmune condition where a virus (dead or alive) triggers the body to form antibodies, which mistakenly attack the myelin sheaths of the peripheral nervous system

s/s: acute, rapid ascending sensory and motor deficit that may stop at any level of CNS, protracted regression, paresthesias, symmetrical motor losses

tx: aggressive respiratory care, tx psychosocial

5. Autoimmune Thrombocytopenia

s/s: small, flat pinpoint micro hemorrhages (petechiae), purpura, prolonged bleeding after venipuncture, weakness, fainting

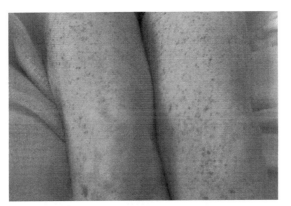

(image credit: bloodspecialistclinic.com)

labs: low platelets count, high PT/INR, decreased Hgb and Hct
nsg: platelet transfusions, corticosteroids, plasma infusion, prevent/control hemorrhage, fall precautions

6. AIDS

dx: ELISA, Western blot assay, low CBC counts, low CD4 counts
s/s: oral thrush, headache, aseptic meningitis, peripheral neuropathies, cranial nerve palsy
nsg: HAART therapy, high protein and calories diet, provide support
Opportunistic infections: jiroveci pneumonia, candida, cytomegalovirus, Kaposi's sarcoma, AIDS dementia complex

Amphotericin B	Antifungal, systemic	S/E: GI upset, hypokalemia, seizures, hematological, cardiac, and hepatic abnorms, protect med from sunlight
		-cannot give within 1 hr of receiving blood
Zidovudine (AZT)	Antiviral, used mainly with AIDS	S/E: Anemia, paresthesias, agranulocytosis
		Teach pts to comply strictly with dosing schedule

Autoimmune Disorders Quiz

1. What are the three largest problems in multiple sclerosis?

2. What are the two main treatments of lupus?

3. Thrombocytopenia will show what in labwork?

4. What is the most important part of treating a Guillain-Barre patient?

5. What is a myasthenic crisis?

6. How are Multiple Sclerosis and Guillain-Barre similar? How are they different?

Psychiatric

1. Anxiety

<u>Mild</u>- high degree of alertness, mild uneasiness, butterflies in stomach

<u>Moderate</u>- incr perspiration, light-headedness, muscle tension, urinary frequency, nausea, anorexia, diarrhea, increased BP, dry mouth, selective inattention, poor comprehension

<u>Severe</u>- hyperventilation, dizziness, vomiting, tachycardia, panic, inability to hear or speak, further decr perception, hallucinations, delusions

<u>Panic</u>- symptoms of severe anxiety, inability to function, dread, terror, personality disorganization

All anti-anxiety medications (except Buspirone) have the potential for abuse and will all cause drowsiness, lethargy, sedation, constipation, etc.

Diazepam/ Chlordiazepoxide	Antianxiety medication	S/E: Lethargy
Alprazolam	Antianxiety medication	S/E: Confusion, check renal/hepatic function
Midazolam	Antianxiety medication used pre-op and post-op	S/E: Retrograde amnesia
Buspirone	Antianxiety medication	Requires 3 weeks to be effective, cannot be given PRN

2. Phobias

Avoid confrontation and humiliation, do not focus on getting patient to stop being afraid, systematic desensitization by trained pros only, general anxiety measures, may be managed with antidepressants

3. Anorexia nervosa v. Bulimia

<u>Anorexia</u>- fear of obesity, dramatic weight loss, distorted body image, anemia, amenorrhea, cathartics and enemas may be used for purging, excessive exercise, vomiting

<u>Cx</u>: may have cardiotoxicity, electrolyte imbalance

<u>Nrsg</u>: **monitor hydration and electrolytes,** behavior modification may help, support efforts to take responsibility for self, explore issues regarding sexuality, goal is to gain ½ -1 lb per week

<u>Bulimia</u>- characteristics of anorexia + binge eating, may be of norm weight or overweight

<u>Nrsg</u>: May be managed with antidepressants, **monitor hydration and electrolytes**

BMI - Body Mass Index - not as accurate in heavily muscled individuals

$$BMI = weight (lbs) \times 703 / height (in)^2$$

< 18.5 = underweight

18.5 - 24.9 = normal

25.0 - 29.9 = overweight

> 30.0 = obese

4. Obsessive-Compulsive Disorder

<u>s/s</u>: rituals or procedures the pt feels s/he must follow, serves as a poor coping mechanism to manage anxiety

<u>Nrsg</u>: accept ritualistic behavior, structure environment, provide for physical needs, offer alternative activities, **guide decisions, minimize choices**, encourage socialization, group therapy, stimulus-response prevention

<u>Tx</u>: may be treated with antidepressants, mood stabilizers

5. Conversion Disorder- physical symptoms with no organic basis, unconscious behavior- could include blindness, paralysis, convulsions without loss of consciousness

-often has a lack of concern about symptoms

<u>nrsg</u>: diagnostic eval, discuss feelings instead of symptoms, promote therapeutic relationship with patient

6. Defense Mechanisms

<u>Denial</u>- ignoring or blocking external events from awareness

<u>Displacement</u>- redirection of feelings to subject that is acceptable or less threatening

<u>Dissociation</u>- person loses track of time, self, and reality and seems to view themselves from an outside perspective

<u>Rationalization</u>- putting something into a different light or changing perspective in the face of reality

<u>Regression</u>- reverting to an earlier stage of development

<u>Projection</u>- attribution to others of one's own unacceptable thoughts, feeling, qualities

<u>Symbolization</u>- less threatening object is used to represent another (eg, phobia of elevators)

<u>Substitution</u>- Replacing desired, impractical or unattainable object with one that is attainable

<u>Repression</u>- unacceptable thoughts kept from awareness

<u>Reaction formation</u>- expressing attitude directly opposite to unconscious ones

7. Depression- low self-esteem, feelings of helplessness/hopelessness, obsessive thoughts, sense of doom, regressed behavior- immature, demanding

<u>nrsg</u>: Watch for clues of impending suicide- any sudden change in patient's behavior/attitude, finalizes business or personal affairs, gives away valuable possessions or pets, withdraws from social plans, has death plan, makes indirect statements ("I may not be around")

-2 week pt during drug therapy most dangerous- have enough energy to carry out plan

-try to **guide decisions and minimize choices** due to low energy

**All anticonvulsant, antidepressant, and antipsychotic medications have the potential to lower the CBC and reduce WBCs, RBCs, and platelets. **

Phenelzine, Isocarboxazid	Antidepressants, MAOIs	S/E: HTN crisis when taken with foods containing tyramine (aged cheeses, meats, wines), do not take with SSRIs/CNS deps
Fluoxetine, Sertraline, Paroxetine	Antidepressants, SSRIs	S/E: Dry mouth, sexual dysfunction, weight gain Take in AM for 4 FULL WEEKS -monitor for decr in CBC values -no Digoxin
Venlafaxine, Duloxetine	Antidepressants, SNRIs	S/E: Anxiety, restlessness, dry mouth, unusual thoughts or behaviors
Amitriptyline	Antidepressants, Tricyclic	S/E: Photosensitivity, bone marrow depression Take at night, 1-3 weeks for therapeutic eff

8. Electroconvulsive Therapy Guidelines

pre: prepare patient by warning about temporary memory loss, get informed consent, NPO after midnight for early morning procedure, have pt void, remove jewelry, dentures, glasses, give muscle relaxant and barbiturate to induce short-term anesthesia

post: take vital signs, orient patient, observe pt's reaction, observe for indications of suicide immediately after ECT treatment, back pain is NOT a normal SE

9. Bipolar Disorder- manic phase + depressive phase

Depressive phase- just like depression

Manic phase- disoriented, euphoria, delusions of grandeur, flight of ideas, lacks inhibition, talks excessively, can't stop moving to eat, weight loss, hypersexual

Nrsg: simplify environment, decr stimuli, finger foods, limit people, distract and redirect energy, set limits, communicate with firm approach, encourage rest

**All anticonvulsant, antidepressant, and antipsychotic medications have the potential to lower the CBC and reduce WBCs, RBCs, and platelets. **

Lithium	Mood stabilizer Blood level: 1-1.5 meQ	S/E: Fine hand tremors, reversible leukocytosis Intoxication: Coarse hand tremors, vomiting, diarrhea, ataxia, drowsiness *Have good salt/fluid intake*

Carbamazepine	Mood stabilizer, anticonvulsant	S/E: Drowsiness, ataxia, decr in CBC values

10. Schizophrenia- altered thought processes, withdrawal from the world and reality

S/s: inappropriate behavior such as silly laughing, transient hallucinations, sudden onset of mutism, bizarre mannerisms, waxy flexibility, suspicion and ideas of persecution and delusions and hallucinations

Nrsg: Do not argue with patient or reinforce hallucinations, communicate in calm and authoritative manner, avoid direct questions, accept client's indifference and avoidance behavior, verbalize what you observe client doing, provide simple activities to incr self-esteem, use honesty and genuineness with paranoid patients

Haloperidol	Antipsychotic	S/E: Low sedative effect, high incidence of EPS
Chlorpromazine	Antipsychotic	S/E: High sedative effect, hypotension, high incidence of EPS
Risperidone	Antipsychotic	S/E: Significant weight gain, mild EPS
Aripiprazole	Atypical antipsychotic	S/E: Weight gain, blurred vision, nausea and vomiting

**EPS: Pseudoparkinsonism, dystonia (muscle spasm), acute dystonic reaction (tightening of jaw, oculogyric crisis), akathisia, tardive dyskinesia

*Neuroleptic malignant syndrome- rigidity, fever, ANS dysfunction, seizures, coma

tx: IM benadryl, cogentin

Diphenhydramine	Blocks effects of histamines (allergic rx) and also reduces S/E of psych medications	S/E: Drowsiness, confusion
Cogentin	Reduces EPS S/E of psych meds	S/E: Drowsiness

11. Alcohol Abuse

nrsg: watch for symptoms of withdrawal (increasing vital signs, tremors, vomiting, diarrhea, SEIZURES), encourage support groups
-most important question: "When did you have your last drink?"
-if hallucinating while withdrawing, turn the lights on in the room
-do not give coffee to a pt suffering from withdrawal or to sober up
-delirium tremens (DTs) may occur 2-3 days after cessation

-best way to talk to a pt about alcohol abuse is to point out the consequences

Disulfiram	Anti-abuse	S/E: Accidental alcohol intake may cause flushing, nau/vom, throbbing headache, syncope
		Avoid: mouthwash, furniture polish, cough syrups, perfume, colognes

<u>Wernicke's Encephalopathy</u>- nutritional deficiency of thiamine common in alcoholism
s/s: ophthalmoplegia, ataxia, confusion
tx: thiamine supplementation, often complete cure

<u>Korsakoff's Psychosis</u>- nutritional deficiency of thiamine that tends to follow Wernicke's
and results in permanent brain damage
s/s: inability to form new memories, loss of memory, confabulation, hallucinations

12. Post-Traumatic Stress Disorder (PTSD)

<u>nrsg</u>: encourage client to talk about painful stored memories with as much detail as
possible, use empathic responses to the expressed distress, remain nonjudgmental, point
out irrational thinking, help client recognize the limits of his/her control over the event,
involve client in anger management program, run a regular exercise schedule with the client

-flashbacks- do not try to speak with/touch patient in mid-flashback

13. Dissociative Identity Disorder (DID)

-formerly known as Multiple Personality Disorder

s/s: features at least two distinct and relatively enduring identities, memory impairment

-poor prognosis

-many sufferers report childhood sexual and/or physical abuse

14. Personality Disorders

<u>Antisocial</u>- "psychopath" - may antagonize, manipulate, or treat others with indifference.
May break the law without remorse, torture or kill animals, and violate the rights of others.

<u>Borderline</u>- brief and intense relationships, temper tantrums, manipulative, repetitive self-
mutilation, depression, labile mood, blames others for own problems

<u>Narcissistic</u>- arrogant, appears indifferent to criticism, sense of entitlement, uses others to
meet their own needs

<u>Dependent</u>- passive, problems working independently, anxious or helpless when alone

<u>Avoidant</u>- timid, social fear, withdrawn, sensitive to criticism, lacks self-confidence

<u>Obsessive-compulsive</u>- sets high personal standard for self and others, preoccupied with
rules, lists, and details, rigid and stubborn, cold affect, repetitive thoughts/actions

<u>Nrsg</u>: limit setting, do not allow behaviors that threaten rights of others, avoid arguing, be
alert for potential manipulation, help verbalize feelings, very little chance for "recovery" or
improvement

Psychology Quiz

1. In panic levels of anxiety, perception is increased. T/F?

2. General guidelines of phobia treatment?

3. Biggest medical complication of anorexia?

4. Treatment of OCD includes what therapeutic guidelines by the nurse?

5. Treatment of conversion disorder by the nurse?

6. What is reaction formation?

7. Describe signs of impending suicide.

8. Describe the prep for ECT.

9. Diet for a bipolar patient who is currently manic?

10. How to deal with hallucinations?

11. Increasing VS in alcoholics going through withdrawal increases the risk of...

12. The most important aspect to treatment of personality disorders is...

Herbs

ALL HERBS INCREASE BLEEDING. THEY ALL CAUSE BLEEDING. BLEEDING BLEEDING BLEEDING BLEEDING BLEEDING BLEEDING BLEEDING BLEEDING BLEEDING BLEEDING

Ginkoba/Ginko	Memory loss, depression, circulation	<u>do NOT take with</u>: MAOIs, anticoags, antiplatelets, cephalosporins, antiseizure meds b/c prolongs bleeding times
Ginseng	Inflammation	<u>do NOT take with</u>: estrogen, corticosteroids, anticoagulants, NSAIDS
Echinacea	Colds, fevers, UTIs	<u>do NOT take with</u>: immunosuppressive agents, methotrexate, ketoconazole, HIV/AIDS, pregnancy
Kava-kava	Insomnia, mild muscle aches and pains	<u>do NOT use with</u>: CNS suppressants, levodopa, MAOIs
St. John's Wort	Depression (mild or mod ONLY)	<u>do NOT use with</u>: alcohol or antidepressant meds (will incr adverse effects), digoxin, opiates
Ma Huang	Asthma, hay fever	<u>do NOT use with</u>: MAOIs, sympathomimetics, theophylline, Digoxin
Garlic pills	Blood thinning, hypercholesterolemia	<u>do NOT take with</u>: insulin/metformin (incr eff), anticoags
Black Cohosh	Hot flashes by creating estrogen-like activity	<u>tx</u>: do NOT take with: antihypertensives, hypotension, breast CA
Feverfew	Migraines, arthritis, fever, dermatitis	<u>do NOT take with</u>: anticoagulants, thrombolytics b/c prolongs bleeding time
Licorice Root	Stomach ulcers, heartburn, colitis	<u>S/E</u>: low potassium
Ginger	Antiemetic, anorexia, motion sickness	<u>do NOT take with</u>: bilberry, feverfew, garlic, ginkgo, other anticoags
Hawthorn	High BP, weak heart muscle	<u>do NOT take with</u>: antihypertensives, nitrates
Saw Palmetto	BPH	<u>do NOT take with</u>: iron pills
Valerian Root	Anxiety, insomnia	<u>do NOT take with</u>: sedatives, antidepressants, decr hepatic func

Herbs Review Quiz

1. What herb should NOT be used with antidepressants?

2. Echinacea should NOT be taken if…

3. Black Cohosh treats what medical condition?

4. Ginger should not be taken with…

5. Ginkoba is generally taken for what purpose?

Extra Resources

All material that follows is in addition to my tutoring course material. It's intended to be just that little bit of extra help you need to ROCK your NCLEX!

Delegation Guidelines

Registered Nurse will treat patients who:

-show signs of a complication or emergency

-are being discharged and need teaching

-need monitoring of TPN, central line care, CVP measurements, extensive dressing/wound care, and surgical drains

-are receiving blood transfusions

-experience fluctuating vital signs

-present with a newly onset problem (eg, confusion)

Licensed Practical Nurse will treat patients who:

-are already stable, in stable situations (eg, patient with fractured femur repaired 4 days ago)

-need to be kept safe with side rails, restraints

-have specimens to be collected

-need oral suctioning

OK to: IM medications, PO narcotics, insert NG, foley catheterizations

Unlicensed Assistive Personal (UAP) will treat patients who:

-need vital signs taken

-need I&Os recorded

-need assistance with oral feeding

-need assistance ambulating as ordered by MD

Common Lab Values

Hemoglobin: M- 13.5 – 16.5

         F – 12.0 to 15.0

Hematocrit: M- 41 to 50

         F – 36 to 44

Platelet count: 100,000 to 450,000

Calcium: 8.0 to 10.0

Magnesium: 1.6 to 2.4

Potassium: 3.5 to 5.0

Sodium: 135 to 145

Glucose 70 - 100

Hemoglobin A1C: 3-6% is normal

         >6% means bad diabetic control

AST/ALT: <35

WBC: 4,500 to 10,000

BUN: 5 to 20

Creat: 0.7 to 1.3

Albumin: 3.5 to 5.0

Total cholesterol: <200

Triglycerides: <150

LDH: <100

HDL: (men) >40

HDL: (women) > 50

INR: 1.0 to 1.2 - pt not anticoagulated

INR: 2.0 to 3.0 - typical anticoagulated range

Cultural/Racial Hints for NCLEX

Judaism:

-pain control during end of life is very important

-meal trays must be Kosher and cannot combine meat and milk

Roman Catholic:

-wafer known as the Eucharist is offered on Sundays/end of life situations

Islam:

-end of life patients should face Mecca

Christian:

-anointing forehead with oil is common

Chinese/Asian:

-may avoid eye contact with authorities to show respect

-may refuse ice water during labor and ask for hot water instead

-criticism or disagreement is not expressed; head nodding does not necessarily mean agreement

Hispanic:

-admiring a child during the initial encounter with a stranger should be avoided, because this may afflict the child with the "evil eye"

-may use dramatic body language to express emotion or pain

-direct confrontation disrespectful

African American:

-do not ask personal questions during initial contact; this may be viewed as intrusive

-beta blockers are not as effective as ace inhibitors for this group

Native American

-silence indicates respect for the speaker

-obtaining input from extended family important

-eye contact may be viewed as a sign of disrespect

-expect to arrive late to appointments

-interpreter etiquette: look directly at patient when speaking, keep questions short and simple, do not interrupt

Legal Considerations

Code of Ethics for Nurses- developed as a guide for carrying out nursing responsibilities in a manner consistent with quality in nursing care and the ethical obligation of the profession, practice with compassion/respect

1990 Patient Self-Determination Act- requires many hospitals, nursing homes, home health agencies, hospice providers, health maintenance organizations, and other health care institutions to provide information about advance directives

Pt's Bill of Rights- client is responsible for providing information about their medications and past history, pt has the right to an easy to understand summary, young adults can stay on a parent's healthcare plan until they are 26, etc.

Patient Protection & Affordable Care Act- free preventative care, covers pre-existing conditions, Rx discounts for seniors, health insurance marketplace

2012 National Patient Safety Goal- minimum of two identifiers, never use room #

Tarasoff Act - duty to warn, duty to protect psychiatric health care providers who are threatened

Autonomy- the right of individuals to make decisions for themselves

Beneficence- A nurse's duty to do what is in the best interests of the client

Justice- a fair, equitable and appropriate treatment

Nonmaleficence- a nurse's duty to do no harm

Fidelity- keeping faithful to ethical principles and the ANA Code of Ethics for Nurses

Virtues- compassion, trustworthiness, integrity, and veracity (truthfulness)

Confidentiality- maintaining the client's privacy by not disclosing personal information about the client

Accountability- responsibility for one's actions

<u>Assault</u>- verbalization of intent to cause physical harm

> Ex: "If you don't take these pills, I will shove them down your throat!"

<u>Battery</u>- physically touching a patient against their will

> Ex: Applying restraints to a patient without physician's orders

<u>Negligence</u>- the doing of something which a reasonably prudent person would do, or the failure to do something which a reasonably prudent person would do

Ex: A nurse checks pedal pulses for the first time on a patient who returned from a cardiac catheterization four hours ago

<u>Slander</u>- words falsely spoken that damage the reputation of another

> Ex: A nurse tells a patient that a doctor is incompetent (with no evidence)

<u>Advance Directive</u>- a document stating a person's wishes about health care when that person cannot make his or her own decisions

<u>Emancipated minor</u>- a person younger than 18 years of age who lives independently, is totally self-supporting, is married or divorced, is a parent even if not married, or is in the military and possesses decision-making rights

*note: any emancipated minor OR someone over 18 may refuse medical treatment (ie: ambulances)

<u>Consent Forms</u>: The nurse…

-may NOT get the patient to sign the consent form (physician will)

-may NOT initially educate the patient about the procedure (physician will)

-may stand as witness when the consent form is signed

-should check that the consent form is signed before surgery

-may answer any follow-up questions from the patient about the procedure

<u>Abuse</u>:

Immediately report upon suspicion: vulnerable patient populations (children younger than 18, elderly, mentally disabled, anyone who cannot advocate for themselves)

DO NOT REPORT FULLY CAPABLE ADULTS.

-when suspecting abuse, get the patient alone ("I need a urine specimen in the presence of a nurse") and ask "Are you being abused?"

-tell abuse victim to hide an overnight bag with clothes and important documents

-If it is discovered that a patient is an illegal alien or cannot pay for treatment: do nothing, continue care.

-Do not tell a fellow nurse details about a patient unless she/he is directly involved in patient's care.

<u>Evidentiary Specimen Collection</u> (ie, for rape victim in ER)

-wrap bullets, shotgun wadding in gauze and put in a cup or envelope

-save any gravel, soil, grass, twigs or glass that are on the victim or on the sheets used for transport

-swabs of both dry and moist secretions should be air-dried prior to placement in appropriate container

-bag victim's clothing in PAPER bags, not plastic

Nursing Management Guidelines

-when you need to report something, ALWAYS go up the chain of command (usually to nursing supervisor)

-if charge nurse resolving issue, always involve staff in seeking solutions to have a more permanent effect

-do not aggressively confront staff

-when correcting nursing assistants, do not be judgmental. Pick the answer that educates.

-if a client asks for unlimited supplies right before they leave, ask "What will you need in the next hour?"

-if staff members are arguing with each other, schedule a meeting with both of them. Discuss all issues and concerns, including negative comments. Frequent exchange of feedback is better, choose the answer where each party is actively listening and hears the other person's perspective

-crisis management: autocratic or directive leadership is best

Nursing Process

-RNs must reassess when pt is refusing pain med, pain med doesn't work, or pt has an option of pain meds

-SBAR - typical format for a nursing note- stands for situation, background, assessment, recommendation

Incident reports- never on chart or mentioned in note

       -send to supervisor, risk management, state what you saw, assessment complete, no injury noted, VSS, MD notified

Typical ER Patients

-"A patient tries to give away his expensive watch to the nurse" – suicidal

-"A pregnant woman in the first trimester (1-12 weeks) with painless spotting" – okay

-"A pregnant woman in the first trimester with unilateral pain and spotting" – ectopic pregnancy

-"Lethargic 2-month-old who has refused to nurse for the past 8 hours" – dehydration, urgent

-"50-year-old male complaining of nausea and is diaphoretic" – heart attack

Precautions Lists

Sequence for Donning PPE- Gown first, mask or respirator, goggles or face shield, gloves

Sequence for Removing PPE- Gloves, face shield or goggles, gown, mask/respirator

Airborne (**My Small Chicken Has Tb**)

Measles (rubeola)

Smallpox / SARS

Chicken pox

Herpes zoster

Tuberculosis

Management: Private room with negative airflow pressure, minimum of 6-12 air exchanges per hour, UV germicide irradiation, filter is used, N95 mask for TB

Droplet (**SPIDERMAn**)	Contact (**MRS. WEE**)
Scarlet fever	**M**ulti-resistant organism
Sepsis	**R**espiratory Syncytial Virus
Streptococcal pharyngitis	**S**kin Infections
Parvovirus	
Pneumonia	**W**ound Infection
Influenza	**E**nteric Infection (Cdiff)
Diphtheria	**E**ye Infection (Conjunctivitis)
Epiglottis	
Rubella	
Mumps	
Mycoplasma	
Ade**N**ovirus	

50 Most Common NCLEX Question Topics

1. Coumadin- check PT, antidote is vitamin K.
   Heparin- check PTT, antidote is protamine sulfate.

2. Gaits for use during crutches

3. Study all the special diets/foods flashcards

4. Study all the developmental milestones

5. High-pitched cry, lethargy, and bulging fontanelles are signs of increased ICP in infants

6. Nurse- unexpected outcomes, first assessments, teaching
   LPN – taking care of med/surg
   Aide – Baths, blood sugar checks, recording I & Os

7. von Willebrand's – no NSAIDs or aspirin

8. Chest tube vigorous bubbling- air leak

9. Chest tube pulls out of person- occlusive dressing taped on three sides
   Chest tube pulls off of container- put end in sterile water

10. Suctioning guidelines- do nose before mouth, no more than 10 seconds

11. Study the Glasgow coma scale.

12. Suicide questions- watch for ambiguous "I may not be around" statements or where the patient is giving valuable/cherished items away OR a sudden change in energy levels.

13. When a patient is taking Lithium, make sure they have plenty of water and sodium in their diet. Proper lithium level is 1- 1.5.

14. For a blood clot, treatment is bedrest.

15. Arterial insufficiency- Cool foot, thin, shiny, atrophic skin, pallorous toes, cyanotic nail bed, absent pulse, round ulcerations of heels and toes.
    Venous insufficiency- Rubra, coarse thickened skin, varicose veins, brawny, pulse present, irregular ulcerations on bony prominences (ankle and heel)

16. The Birth Control Pill puts patients at risk for blood clots, especially if they are SMOKING.

17. Digoxin- Norm levels are 1.0 to 2.0. Low potassium increases risk for toxicity. Toxicity signs are nausea, vomiting, confusion, drowsiness and halos around lights.

18. Give the bronchodilator medication and THEN the longterm medicine.

19. Signs of fluid overload: cool skin, crackling, cyanosis, edema.

20. Memorize the different types of blood transfusion reactions.

21. Wounds most likely to dehisce? Abdominal wounds. What to do? Put on sterile gauze soaked in saline and call doctor immediately.

22. Signs of compartment syndrome- pulselessness, pallor, paresthesias, pain, pressure.

23. Autonomic dysreflexia- immediately raise HOB and decompress bladder, give anti-HTNs

24. AIDS precautions are standard, NOT contact.

25. When there are STDs in your pt who keeps getting reinfected, make sure the partner is getting treated as well.

26. For a late deceleration in the HR of a fetus, turn the mother on the left side, give oxygen, stop pitocin.

27. Always support efforts to go to outpatient support groups. They will do more for the patient than educational materials post-hospital.

28. Fix cleft lip 2-3 months after birth, cleft palate 12-13 months after birth

29. Understand how to calculate Apgar scores.

30. Know how to perform medication calculations.

31. Treatment of ventricular fibrillation- immediate emergency shock and CPR. Treatment of cardioversion- give drugs, sedate patient, sync rhythm, controlled shock.

32. Where to look for jaundice in dark-skinned people? In the gums.

33. Phototherapy for babies- put on sunglasses/eyepatches. Tx for jaundice.

34. Excessive bleeding for chest tubes is >100mL/hr.

35. Initial antidepressant use- watch for sudden change in mood(suicide

36. Be able to recognize EPS effects in clients taking antipsychotic medications, especially neuroleptic malignant syndrome.

37. Pregnant/breast-feeding mothers' breasts: supportive, well-fitting bras, don't wash with soap, only water, apply warm compresses if breastfeeding and have clogged duct. Apply ice & avoid nipple stimulation if mother does not want to breast-feed

38. Personality comes from the frontal lobe.

39. Elder and child abuse should be reported upon suspicion; capable adults should NOT be reported to any agency unless they ask for it.

40. Blood sugar hypoglycemia- sweaty, cold, clammy, confused
Hyperglycemia- hot dry skin, confusion

41. TB drug side effects- peripheral neuropathy- given Vitamin B6 as a prophylaxis for this effect. Also, watch liver function tests because they're tough on the liver. Rifampin turns tears, urine, sweat orange.

42. To draw an ABG, first perform an Allen's test. Know how to read the results.

43. To test for carpal tunnel syndrome, press the back of the hands against each other, bending the wrists, and observe for sharp pain.

44. Traction should never be removed, taken off the bed. Never prop their knees up, but elevating foot of bed on "bed blocks" is okay.

45. Patients with seizure precautions should be on a mattress on the floor, no padded side rails.

46. TPN side effects- do not quickly increase or decrease rate to avoid hypo/hyperglycemia and/or fluid overload. TPN expires after 24 hours.

47. CPR- Rate of 100 compressions per minute in a 30:2 ratio in adults, 3:1 in infants with one rescuer, 15:1 in infants with two rescuers.

48. Know pediatric milestones.

49. Know pediatric immunization schedules.

50. Dye for CT scan and cardiac catheterization- allergy to shellfish, Flu shots- allergy to eggs.

My NCLEX Plan

Test Date & Time:

Address of the Testing Facility:

Physical exercise to do the day/evening before?

Bed time relaxation treat?

Do you know how to get there and how long it will take?

What time will you wake up so that you have time for breakfast?

What time will you leave the house?

Positive phrase you will say to yourself in your car?

Positive song in the car to get you pumped?

Reward after the test?

Tips and Techniques For NCLEX

- Try to get in some physical exercise the day/evening before the test. That will help you to sleep even if you're anxious.

- Set two alarms to ensure you get up.

- Leave enough time to eat breakfast.

- Use the earplugs.

- Remember that other people are in the testing center for things other than NCLEX. If they leave before you, don't panic! They're probably taking a different test.

- Take your time and read each question TWICE.

- Note whether the question is asking for the INCORRECT or the CORRECT answer.

- If you are in long enough to get to a break and you need to pee/get a drink, DO IT.

- Try one of the in-test physical relaxation techniques below.

- Trust what you know and rely on your intuition. Don't pick an answer as correct just because you've never heard it before.

- Engage in positive thinking! Tell yourself "I can do this, I just need to concentrate. I've been studying a long time for this, and I've gotten so much better. I will do the best that I can with the questions I am asked. This is NCLEX: I don't have to get every question right, and not every question is counting toward my score.

In-Test Physical Relaxation Techniques
(pick one and stick with it!)

Tense and Relax Method

1. Put your feet flat on the floor.
2. With your hands, grab underneath the chair.
3. Push down with your feet and pull up on your chair at the same time for about five seconds.
4. Relax for five to ten seconds.

The Beach Method

1. Close and cover your eyes using the center of the palms of your hands.
2. Prevent your hands from touching your eyes by resting the lower parts of your palms on your cheekbones and placing your fingers on your forehead. Your eyeballs must not be touched, rubbed or handled in any way.
3. Think of some real or imaginary relaxing scene. Mentally visualize this scene. Picture the scene as if you were actually there.
4. Visualize this relaxing scene for one to two minutes.

The Deep Breaths Method

1. Sit straight up in your chair in a good posture position.
2. Slowly inhale through your nose.
3. As you inhale, first fill the lower section of your lungs and work your way up to the upper part of your lungs.
4. Hold your breath for a few seconds.
5. Exhale slowly through your mouth.
6. Wait a few seconds and repeat the cycle.

THE END

You're reached the end of the course. Yay!

Final advice:

1. Work on questions every day. Focus on a recommended source (see NCLEXSimplified.com) or subscribe to the exclusive access CORE questions. Be efficient!

2. Don't take the test until you're scoring 75% or higher regularly on questions.

3. Take a deep breath and good luck!

4. After you pass, remember to leave me a review so others know that my course works! The NCLEX Simplified Facebook page is a good place to do so. :)